50 Reproducible Activities for Building Innovation

Richard Brynteson, Ph.D.

HRD Press, Inc. ❖ Amherst, Massachusetts

Published by: HRD Press
22 Amherst Road
Amherst, MA 01002
800-822-2801 (U.S. and Canada)
413-253-3488
413-253-3490 (fax)
www.hrdpress.com

ISBN 978-1-61014-388-2

Production services by Jean Miller
Cover design by Eileen Klockars
Editorial work by Sally Farnham

Table of Contents

Expanded Table of Exercises

PART 3

PART 4

Why innovation now?

President Obama is talking about innovation. Governor Dayton of Minnesota as well as other governors are talking about innovation. Corporate managers are asking their employees for more innovation. People are studying the life of the deceased Apple founder Steve Jobs and asking why more of us can't be more like him. Where have you gone Thomas Edison and Leonardo da Vinci? How do we tap in to the human spirit and mental resources to become more innovative?

The western world seems to be enmeshed in a three-year recession. It might even be a fundamental economic restructuring. Jobs have been lost; some industries are sputtering on life support. Even industries where the United States has shown supremacy, like education and medicine, are taking competitive hits from Asia. The way out of this recession is not more spending on health care or government services; it is innovative thinking. It is developing new and better products. It is creating efficient processes for work. It is embracing a spirit of innovation that states unequivocally: we can do better, we can be better, and we can think better.

Purpose of This Book

The purpose of this book is to create centers of innovation. We are past the point where we can rely on brilliant or creative individuals. Innovation has to be organization-wide. Organizations have to build capacity for innovation so that they can produce innovation after innovation. This book presents ways to build that capacity. This book provides exercises and activities to build the innovation muscle of individuals, groups, and organizations.

It is my strong belief that innovation can be a learned trait by individuals, groups, and organizations. These exercises provide a road map, a method, an impetus to develop that trait.

How and Where to Use

These exercises do not need to be used in any particular order. The time parameters do not need to be followed. The procedures do not have to be followed verbatim. The exercises can be plucked and harvested in any number of ways.

Some situations where these exercises may be used include the following:

- for standalone play during "lunch and learn" sessions at your organization
- in organization-wide innovation training sessions
- strategically placed during process redesign sessions in order to shake up the thinking of participants
- peppered into fun events, such as scavenger hunts and company picnics

- in problem solving, when tackling organizational problems head-on
- as team-building activities for teams and groups
- as activities for friendly competition between work groups

In all cases, these exercises will help build the organizational innovation muscle.

Why me?

In my late 20s, while working for a chemical company, I participated in a brainstorming session. Another participant later noted that I contributed about half the ideas that emerged that particular day. Other people had deemed me creative but I had brushed off those compliments. I began to research this concept and later in the 1990s wrote my dissertation on the creative process. I continued to be intrigued by the question "How does creativity happen?"

In the late 1990s, I shifted my focus from creativity to innovation. Innovation had more cache than creativity in organizations. My research question shifted to "How can organizations become more innovative?" I took nine trips to Asia in the last decade, working primarily with the military organizations on innovation efforts. We worked on projects with F-16 fighters, Chinook helicopters, and navy patrol boats. If an entrenched military can become more innovative, then why can't other organizations? I used many of these exercises in those efforts. Continuing to make organizations more innovative has become my passion for this segment of my life.

INTRODUCTION

So, what is innovation? Wikipedia defines innovation:

> Although the term is broadly used, innovation generally refers to the creation of better or more effective **products, processes, technologies,** or **ideas** that are accepted by **markets, governments,** and **society.** Innovation differs from **invention** or **renovation** in that innovation generally signifies a substantial positive change compared to incremental changes.

This definition is comprehensive and useful.

What innovation is:

- doing something significantly better
- creating a stream of new and useful products
- making something better in order to improve the world, or make money, or both
- streamlining processes so that they are faster, cheaper, better, and more efficient
- revolutionizing entire industries
- facilitating creative change that will "woo" people
- leapfrogging the competition with products they had never thought of

What innovation is not:

- not just about iPods
- not just about products
- not just for scientists and engineers
- not out of reach of civilians, business people, and just us ordinary people
- not just about complex formulas
- not about the lone Leonardo da Vinci or Thomas Edison
- not about abstract brainstorming sessions

This book is a call to action. If you ever thought that innovation was a ball in someone else's court, I hope that the exercises in this book prove you wrong. If you ever put off being innovative because you did not know what to do, you no longer have an excuse. If you ever thought that your organization needs to embrace innovation, now is the time. This book provides a road map.

INNOVATION IS LIKE

Innovation is like

- a seedling growing after a spring rainfall;
- new life emerging after a painfully long labor;
- an upside down traffic light in the midst of a cornfield;
- a drink of fresh, cool water after a long drought;
- a suspension bridge between two disparate ideas;
- an underground tunnel out of imprisoned thinking;
- an eagle that swoops down and grabs a fish from a lake and drops it into a different lake;
- a shooting star that emblazons the sky behind it;
- a pinball bouncing between the creative abrasion of a conversation;
- a concrete manifestation of a wild idea;
- the remnants of two bighorn sheep butting heads on a mountain outcropping;
- a sigh of relief after a long uphill hike;
- a shot in the arm for those at the bottom of the trough.

Questions to Consider

As you embark on this journey of innovation, enter it with an air of inquiry. Consider the following questions (warning: there might not be clear answers):

- Can innovative thinking be a learned behavior, or is it "either you have it or not"?
- Does one have to be brilliant in order to be innovative?
- How can an organization embrace innovation fully and still protect its incumbent businesses?
- Can an organization embrace incremental and breakthrough change at the same time?
- How can organizations keep their trade secrets and embrace "open source" innovation?
- How can organizations build efficient business systems and structures that are still open to breakthrough innovations?
- Can management order innovation to happen?
- Can a department be extraordinarily innovative when embedded in a stodgy organization?

By the way, several of these questions might be powerful conversation starters at the beginning of a strategic planning retreat.

STRUCTURE OF THIS BOOK

This book is broken down into four content areas and several addendums. Following are the principal chapters with exercises:

The Practice of Innovation

These exercises will immerse the participants in practicing to be creative and innovative. These exercises give participants shots in their arms and elevate their senses of innovation.

The Innovative Personality and Skill Sets

These exercises help build individual, group, and organizational innovation skills. They build individual capacity to be innovative.

The Culture of Innovation

It is one thing to develop a lucrative, one-shot innovation. It is another to build a culture that produces innovations on a regular basis. These exercises will help groups and organizations build capacity for ongoing innovation.

The Innovative Process

The process of innovation that is posited by this book contains five steps. This set of exercises shows how to develop these five steps.

The content portion of this workbook is followed by two other sections:

Additional Questions, Posters, and Quotes

This section contains questions to ask during each stage of the innovation process. It also has quotes and sayings that can be posted up in an innovation room.

Cases

This section contains some innovation initiatives (caselets) from the author's consulting experience.

ABOUT THE AUTHOR
Richard Brynteson, Ph.D.

"Richard will push your thinking." As an organizational consultant, executive coach, teacher, and writer, Richard helps his students and clients examine their own thinking and make changes in order to improve their productivity and quality of life.

Richard helps his clients innovate work processes and products in order to improve effectiveness and efficiency. He helps his coaching clients explore the relevance of their own assumptions and paradigms. Richard shines a mirror at his coaching clients so they can see the flaws in their thinking.

Richard has helped the Singapore military redesign work systems around Chinook helicopters, F-16 fighter planes, and navy patrol boats. He has created leadership training programs for several companies, and helped engineers become more personable, executives become less fearful, and managers become more thoughtful.

Representative Clients

Department of Defense, Singapore, Dell Computers, Malaysia, McCann Erickson, Singapore and Kuala Lumpur, Rochester Public Utilities, Austin Utilities, Healtheast, Zumbro Valley Mental Health Center, Firstmark Services, Medtronic, Anoka County

Professional Background

Professor of Organizational Management, Concordia University (current)
Innovation Consultant, Singapore and Kuala Lumpur
Ecolab, Product Line Manager
Software Clearing House, Marketing Manager
W.R. Grace, Marketing Manager and Financial Analyst

> "Pushing the thinking of individuals and groups... innovation is the driving force of the economy and we can all be innovators."

Education

University of Minnesota, Ph.D., Adult Education, 1997
University of Chicago, MBA, Finance and Marketing, 1980
Dartmouth College, BA, English, 1977

Publications

Once Upon a Complex Time: Using Stories to Understand Systems. 2006. www.sparrowhawkmedia.com

The Manager's Pocket Guide to Innovation. 2010, HRD Press, Amherst, MA

The Manager's Pocket Guide to Social Media (with Carol Rinkoff and Jason DeBoer Moran). 2011, HRD Press, Amherst, MA

PART 1
The Practice of Innovation

❖ Forget, Unlearn, Dismantle ❖

🕐 *30 – 45 minutes*

Purpose

The purpose of this exercise is to help participants understand what the first steps in being innovative are.

Materials

- ✓ Flip chart paper
- ✓ Markers
- ✓ Imaginary dynamite

Procedure

1. Introduce the concepts of unlearn/forget/dismantle. This is always the first step in innovation. We must make a space for innovation. We need to let go of the old in order to make way for the new. More importantly, we need to let go of the old concepts that have been guiding our lives in the past.

2. Break the large group into smaller groups.

3. The groups will create one flip chart page (or two or three) that focuses on old concepts that we have let go of as a society. Brainstorm them with the participants and make sure there is a wide range of answers. Do this in order to "prime" the participants. For instance, some answers might include the following:

   a) The world is flat.

   b) Smoking is not bad for us.

   c) The only careers for women are in elementary education, nursing, and administrative positions.

   d) The Soviet Union is going to take over the world.

   e) China is a backward country.

   f) Telephones need cords.

4. Lead a discussion on what this brainstorm tells us.

   a) We may not be right all the time.

   b) Times change.

   c) What was right/appropriate/common knowledge/politically correct at one time may not be so any more.

   d) We can laugh at ourselves and our old concepts about the world.

Forget, Unlearn, Dismantle (concluded)

5. Groups will create several flip chart pages on the wall. This time the topic will be what we can unlearn/forget/dismantle about our organization. (You might want to remind them of a ground rule like confidentiality.) You can prime them with statements like these:

   a) The old billing system works well today.

   b) Our only group of customers is _____ .

   c) The way we develop products is _____ .

   d) "Customer service" is a centralized function in our company.

6. **Optional:** Take one of the brainstormed options and focus on it (with a separate piece of flip chart paper). What actions would it take to dismantle that piece of the organization (or process) or rethink that customer group?

Debrief

- Your goal, as facilitator, is to get the participants to question, if not kill, the sacred cows in their organizations. You need to give them permission to unlearn and forget and dismantle.

- Often, in organizations, employees are stuck in the rut of "it has to be this way." It does not. Many successful organizations re-make themselves constantly in terms of new products, services, processes, and client bases.

- There may be nay-sayers in this group—"we have regulations," "we can't change anything because of corporate." You need to honor their voices while emphasizing what is possible to forget/unlearn/dismantle.

- Ultimately, you are giving participants permission to look hard at all that they have held as "the way it is" in their organization.

❖ What if? ❖

🕐 *45 – 60 minutes*

Purpose

The purpose of this exercise is to engage participants in imaginative thinking. Innovation takes imaginative thinking.

Materials

✓ Flip chart paper
✓ Markers
✓ Wild-eyed imagination

Procedure

1. Reassure the group that "we're playing with ideas" in this exercise. Let them know that there are no wrong ideas, just interesting and engaging concepts.

2. Write "What if?" on several pieces of flip chart paper and invite the group to develop fanciful ideas. Start with more general concepts.

3. You might need to prime the group. For instance,

   a) What if half the days of the year were totally dark and half totally light?

   b) What if cars needed refueling every 10,000 miles only?

   c) What if you had to cut your food budget in half?

   d) What if you had to take in four foster children next week?

   e) What if the Internet went dead for one week?

   f) What if gasoline cost $7 a gallon?

   g) What if water cost $3 a gallon?

4. After 5–10 minutes of this fanciful thinking, choose one or two, and create a flip chart page for that one (or two).

5. Have the participants brainstorm the *implications* of that "What if?" Again, they can be fanciful. The wilder the answers, the better.

6. Do the same exercise for issues involving the participants' organizations. Do the "What if?" thinking and follow it with exploring the implications of one or two of the possibilities.

Debrief

This is a fanciful exercise with no right or wrong answers. A debrief question might be, "What 'what ifs' do we have in this organization?" Where are we too satisfied and happy with *what is* rather than *what could be?*

❖ Innovative Connections ❖

🕐 *30 – 45 minutes*

Purpose

The purpose of this exercise is to force connections. Making obscure connections promotes innovative thinking.

Materials

- ✓ Flip chart paper
- ✓ Markers
- ✓ Worksheet #1
- ✓ Open imaginations

Procedure

1. Divide the larger group into smaller groups of 4–7 participants.

2. Distribute Worksheet #1 to participants.

3. Ask participants to connect one item from Column #1 to one item in Column #2 and create a new product out of that connection. Have them do this with several of the items.

4. When each group has three or four new products, have them report out with the larger group.

5. List their mini-inventions on flip chart pages while everyone enjoys a good laugh at them.

6. Ask each group to reconvene. Ask them to take two of the inventions from other groups and elaborate on them.

7. After another 10 minutes or so, have a whole-group report-out.

Debrief

The wackier the product, the better. You are not trying to win product awards here; you are trying to inspire and build imaginations. This exercise can be used as a warm-up to more serious innovative thinking.

Or you might bring participants directly into their own situation. Have them connect one of their existing products or services with an underserved market. What connections can they make now?

Worksheet #1
Innovative Connections

Connect one item from Column #1 to one item in Column #2 and create a new product out of that connection.

Column #1	Column #2
clothing washer	sock
coffee maker	machete
tire iron	iPad
trampoline	conveyor belt
cafeteria tray	carrot peeler
whiskey bottle	chip clip
Frisbee	baseball bat
Barbie doll	picture frame
bull whip	Legos
coat hanger	airplane wing
blender	door bell
kitchen chair	dog collar
volleyball net	microphone

❖ Levels of Innovation ❖

🕐 *60 – 90 minutes*

Purpose

The purpose of this exercise is to show how there are several levels of innovation available for any organization.

Materials

✓ Flip chart paper
✓ Markers
✓ Deep thinking
✓ Wild ideas

Procedure

1. Introduce the idea of "levels of innovation" below:

   • Process Improvement Ideas (lean manufacturing, Six Sigma)

   • Derivative Ideas (Starbucks, microloans)

   • Breakthrough Ideas (Harry Potter, space travel)

   • Radical Innovations (iPods, wireless)

2. Hang up eight flip chart pages around the room, two each for each of the four previous levels of innovations. For each level ("Derivative Ideas" for example), add "past examples" for one of the flip chart pages, and "future potential ideas" for the other page.

3. Divide the larger group into four smaller groups. Have each group start at one "level," and brainstorm the past and the future of that level for the organization. Write down all the ideas they can think of that fit that "level" of innovation.

4. Time it for each group to be at each station for 10 to 15 minutes. Ring a bell, blow a whistle, and then tell them to move to the next "level." Do this until each group has spent a chunk of time at each level.

5. Tell participants that some ideas or products or services might fit into more than one of the levels.

6. You might have to put up more flip chart paper as each page fills up with ideas.

Levels of Innovation (concluded)

Debrief
- Ask participants how the process went for them. They probably wanted more time. There should have been much chatter during this period of time.

- Tell participants that there is no rocket science that delineated exactly one level from another one.

- Suggest that organizations should be working at all levels. If there are limited resources, they must make allocation decisions. Regardless of those decisions, many people in an organization must think about what those "breakthrough" ideas might be in this particular industry.

- If you have time, you might look at organizations that are familiar to everyone— Target stores, a sports team, a grocery store chain, Starbucks—and talk about what the various levels of innovation might be in the future.

❖ Alternative Uses ❖

🕐 *10 –20 minutes*

Purpose

The purpose of this exercise is to invite participants to think outside the box.

Materials

✓ Worksheet #2
✓ Bizarre sense of humor

Procedure

1. Divide the larger group into smaller groups of 4 to 7 participants.

2. Distribute Worksheet #2 to participants.

3. Have groups develop lists of broken treasures.

4. Have smaller groups report out to the larger group.

Debrief

This is another fanciful, out-of-the-box exercise. The universe is full of gifts and surprises if we look for them. In third world countries, resourceful people use worn out tires for sandals and discarded bits of wire for shoelaces. They have to be innovative because of their lack of resources. We can learn to build that skill also.

Can you take this exercise further? What are the hidden resources in your organization? What do you discard that may be useful elsewhere? What are hidden treasures in some of the employees in your organization? What skills or insights might you be leaving on the table, discarded or ignored?

Worksheet #2
Broken Treasures

Think about the following "broken" items. What uses can you find for them? What value can you extract out of them? Also, you can combine any of these items.

- Broken cement block

- Torn Grateful Dead T-shirt

- Broken alarm clock

- Flat bike tire

- Broken blender

- Left shoe

- Extra 2" x 4" pieces of lumber

- Old matted feather pillow

- Discarded swing set

- Cracked mahogany salad bowl

- Discarded metal flag pole

- Tattered dog collar

- 10′ x 10′ piece of shag, green carpet

- Sunken aluminum canoe

- Discarded car seat

- 100 outdated, dog-eared books about the Crimean Wars

- 100 pounds of greasy ball bearings

❖ Jobs that Need to be Done ❖

🕐 *30 – 45 minutes*

Purpose

The purpose of this exercise is to help participants think like entrepreneurs. It helps participants look for market opportunities.

Materials

✓ Flip chart paper
✓ Markers
✓ Acute understanding of fellow human beings

Procedure

1. Divide the larger group into smaller groups of 4 to 7 participants.

2. Give each group a target market of consumers that the organization serves. For example, it might be stay-at-home mothers or fathers or teens or retired people or busy young professionals.

3. Ask them to brainstorm the jobs "that the members of the target market need to do."

4. Ask each group to present their findings to the larger group.

5. Have the groups move to the right or left and gather around a different flip chart and elaborate on the findings on that flip chart. Either they can add "more jobs that need to be done" or have them think of products or services that would help get these jobs done.

6. Have the smaller groups report out to the larger group.

7. **Example:** Say your group is stay-at-home mothers or fathers. Your original brainstorm list might look like:

   a) Do the laundry while holding baby
   b) Prepare a meal while entertaining children
   c) Clean the bathroom
   d) Do the grocery shopping
   e) Clean the cupboards
   f) Arrange dental appointments
   g) Entertain toddler while nursing baby
   h) Keep living room clean
   i) Keep car clean and uncluttered
   j) Communicate with spouse

Jobs that Need to be Done (concluded)

Debrief

- Invention is about seeing a need and plugging the hole. First, one has to start seeing the "jobs that need to be done."

- Ask participants to talk about some of their possible inventions. Ask others to elaborate on them. Ask what else is needed to make them marketable.

- Explain to them how cup holders in cars, home delivery groceries, and dog walking services all evolved from such an exercise. Also, think of the new generation of baby strollers. Originally, they were simple. Then they were built with storage areas so that stuff could be carried with the baby. Then they needed to be compact in order to fit into car trunks so they got small again, but still with a storage area or two.

❖ A New TV Program ❖

🕐 *3 – 4 hours*

Purpose

The purpose of this exercise is to make innovation a game and help participants stretch their powers of creation.

Materials

✓ Flip chart paper
✓ Markers
✓ Quirky sense of humor
✓ Laptop computers with Internet access

Procedure

1. Divide the larger group into smaller groups of 4 to 8 participants.

2. Allow each group to choose a demographic/psychographic group:

   a) Tweens
   b) Teens
   c) Young adults
   d) Retired folks
   e) Stay-at-home moms or dads
   f) 18- to 24-year-old angry men
   g) Thirty-something rising professional
   h) Other

3. Have the smaller groups create a blockbuster new TV program for this demographic/psychographic group. Have them create major characters (with names), content, and a blueprint for the first three episodes. (The groups can use the Internet to research any of these subgroups.)

4. After three hours or so, check in with the groups and see if they are ready to report-out to the larger group. They can either use flip chart paper or create a PowerPoint.

Debrief

This is a fun exercise. Typically, participants are fully engaged in this process of creation. During the debrief, the facilitator might ask:

• At which points in the process did you feel fully engaged? Frustrated? Happy?

• How can you relate this exercise to your job?

• Where else do you research a group and then create something for them?

• Having done this exercise, will you look more closely at TV programs and see what they are trying to do and who they are aimed at?

❖ Storytelling ❖

🕐 *30 – 60 minutes*

Purpose The purpose of this exercise is to build the storytelling skill set in participants.

Materials
- ✓ Notebook paper
- ✓ Pen
- ✓ Imagination

Procedure

1. Divide the larger group into smaller groups of 3 to 7 participants.

2. Tell them that their task is to create a compelling story and share with the larger class.

3. A story contains three major parts: action, conflict, transformation. Each of their stories should have these parts.

4. Tell the class some of your favorite stories and tell them why you like them. What makes a story powerful?

5. Each group can choose the subject of their own story. It can pertain to their organization but does not have to. These stories could be the basis for commercials, websites, or brochures. Examples of compelling stories include the following:

    a) The single mother who enrolled in your educational program, built, a career, and pulled her family off of welfare

    b) The recent immigrant who came to the United States penniless and used your social service agency to get a start in life

    c) The husband and wife who were constantly fighting until they bought an appliance that your company manufactures and now have nothing to fight about

    d) The dorky guy who could not get a girl, but now, wearing the suits your store sells, has plenty of social action

6. Have the groups work on the stories until they are reasonably polished.

7. Have them present the stories to the class.

Debrief Lectures are boring; stories are compelling. Ask the class what makes a good story. Ask them what some of their favorite stories are. Dissect these stories with the class. What makes for an interesting story—a story that one wants to keep listening to? Which stories excite us and why?

❖ Love of Failure ❖

🕐 *30 minutes*

Purpose

The purpose of this exercise is to help potential innovators examine their attitudes toward failure.

Materials

✓ Flip chart paper
✓ Markers
✓ Notebook paper
✓ Memories
✓ Worksheet #3

Procedure

1. Divide the larger group into smaller groups of 3 to 7 participants.

2. Distribute Worksheet #3 to participants and allow 5 to 7 minutes for the groups to complete it.

3. Debrief the answers with the larger group. What does this tell you about failure and success? Lead them to the point that failure and success are intertwined.

4. Ask participants to get into smaller groups again.

5. Ask participants to reflect individually on what failure meant in their families as they grew up (for 5 to 10 minutes). Ask them to jot down some notes, documenting some specific examples of their failures and how their parents, teachers, or siblings reacted.

6. Ask participants to share their list of failures and reactions with their smaller groups. One recorder in each group should write down themes that emerge from people's stories.

Debrief

The larger debrief should bring out the themes from each of the groups. The themes should be similar. Some people were punished for failures. Other people were encouraged to learn from them. Others were told to brush them under the carpet. How families deal with failures has a lot to do with how we deal with failure as an adult.

Another part of this debriefing should address how we overcome our past programming and learn to deal with failure happily, successfully, and productively. Innovators need to embrace failure because most innovations are failures, at least at first.

Worksheet #3
Failures of Successful People

Match the successful person/people in Column #1 to the failure in Column #2.

COLUMN #1	COLUMN #2
George Washington	three attempts to find the Northwest Passage
Walt Disney	903 light bulbs that did not work
George Macy	seven bankruptcies
Thomas Edison	Six out of nine battles lost
Abraham Lincoln	47 times did not get off the ground
Wright Brothers	six bankruptcies
Thomas Jefferson	kicked out of the company he founded
Steven Jobs	lost six elections before being elected

Answer Key

COLUMN #1	COLUMN #2
George Washington	Six out of nine battles lost
Walt Disney	six bankruptcies
George Macy	seven bankruptcies
Thomas Edison	903 light bulbs that did not work
Abraham Lincoln	lost six elections before being elected
Wright Brothers	47 times did not get off the ground
Thomas Jefferson	three attempts to find the Northwest Passage
Steven Jobs	kicked out of the company he founded

❖ Visioning ❖

🕐 *1 – 2 hours*

Purpose

The purpose of this exercise is to show the power of visioning and to give participants a concrete example about how to do visioning.

Materials

- ✓ Flip chart paper
- ✓ Markers
- ✓ Notebook paper
- ✓ Imagination (not optional)
- ✓ Computer projector with pictures of various vacant lots and playgrounds (optional)
- ✓ Laptop computers with Internet access (optional)

Procedure

1. Divide the larger group into smaller groups of 3 to 7 participants.

2. Ask participants to close their eyes. Read them this passage. "You (a consultant) have been approached by a neighborhood group with a problem. The city has just purchased a nearby vacant lot. There are no other playgrounds anywhere near this plot of land. The group asked you to develop a physical layout for an ideal playground. In addition, because the city has no money for building the playground, the neighborhood group has asked you to develop a potential list of partners to help bring this project into being."

3. Members of the group might have questions. You do not have answers. They can choose the community and the plot of land.

4. Show slides of vacant lots and playgrounds, if you want. This might help their mental imagining.

5. Let the participants begin the task. Remind them that there are two discrete parts of the task. First, they have to mentally develop and sketch out the physical layout of the playground. Second, they have to create a list of potential partnerships and how these partnerships would fit together to accomplish the task.

6. After 1 to 2 hours, have each group present their plans.

7. **Optional:** The Internet access computers can be used to research potential partners.

Visioning (concluded)

Debrief

"Visioning" and "networking" are two key behaviors of successful innovators. Innovators have to be able to envision an outcome that they are trying to create. They are to have the imagination to envision that which does not exist. Most innovators also have to rely on other resources to bring their innovations to fruition. Who will be those partners? Who has the resources that they do not have? Who will be the angels to fill in the gaps to make this or that project happen?

Possible partners for the playground development project:

- Nearby Home Depot or other building materials store may donate materials.
- PTA group may volunteer labor to help build.
- Trader Joe's usually helps neighborhood groups with projects like this.
- Kiwanis, VFWs, Elks, and other similar clubs are likely to consider giving resources.
- Other neighborhood businesses are likely to support the project.
- Local sports teams are often looking for good publicity.
- Local college and high school students often have to build "service learning" into their curriculum. For instance, many high school students have to perform 40 hours or so of service work.

❖ Green Innovation ❖

🕐 *60 minutes*

Purpose

The purpose of this exercise is to have participants search for innovations within a certain industry—the green industry. Another purpose is to show participants that innovations build on each other.

Materials

- ✓ Flip chart paper
- ✓ Markers
- ✓ Notebook paper
- ✓ Pens
- ✓ Laptop computers with Internet access (optional)

Procedure

1. Divide the larger group into smaller groups. Participants can share computers. Each small group should have a scribe, a sheet of flip chart paper, and a marker.

2. Have each scribe divide the flip chart paper into two columns: **New Products** and **Possible Other Products.**

3. Have participants go to the website treehugger.com, trendhunter.com, or a similar green products–type website.

4. As participants peruse the website, have them call out to their scribe new, interesting products that they are noticing. The scribes should add these products to the first column. Upon hearing this call-out, others can brainstorm what further products might come about that are similar to or as a result of this product. The scribes should add these ideas to the second column.

5. As the sheets fill up, have the participants post them on the walls.

6. After 45 minutes or so, have groups walk around the room and look at the findings of the other small groups.

Debrief

Hunting for trends is one of the first steps in innovation work. Trend hunting should not be limited to a person or two in marketing research. Because of the Internet, anyone can and should be a trend hunter. Besides, trend hunting is fun.

The key phrase in this trend-hunting process is "What might this innovation lead to?" or "What might be the next in line of this line of products?"

You might have a contest and give awards for the most interesting product, most likely to succeed product, and/or the most bizarre product.

❖ Practical Individual Creative Skills ❖

🕐 *30 minutes*

Purpose

The purpose of this exercise is to help individuals assess their own practical creative skills and to create an action plan for building those skills.

Materials

✓ Worksheets #4 and #5
✓ Pens
✓ Insightful self-assessment

Procedure

1. Divide the larger group into smaller groups that are manageable in size. It is preferable that participants be in groups of people who know them well.

2. Distribute Worksheet #4 and Worksheet #5 to participants.

3. Explain to the group the premise and elements of Daniel Pink's *A Whole New Mind* (see Worksheet #4).

4. Tell the group that their task is to critically assess their own skills in the arena of practical creativity. Have group members help assess your capabilities in each of these areas.

5. Have group members help develop action steps to develop your skills in this area. What activities would help you build these skills?

6. Each participant should report to the larger group one or two action items that they intend to take to build these practical creative skills.

Debrief

Daniel Pink's work *A Whole New Mind* is excellent and worth the read. His premise is that many left-brain type jobs can be either sent to India or computerized and therefore employees need to develop their right brain skills in order to survive the marketplace. These skills are listed on the worksheet. These are clearly right brain skills, and sometimes are harder to train for. According to Pink, employees should be trained in these skills more than the usual left brain skills.

Worksheet #4
A Whole New Mind
(Adapted from *A Whole New Mind* by Daniel Pink)

Premise: We should be educating our students more in the right brain. Left-brain jobs will be done by computers or lower-paying locations. Below are six ways to educate youth for the work of tomorrow.

Design

It is no longer sufficient to create a product, a service, an experience, or a lifestyle that is merely functional. Today it is economically crucial and personally rewarding to create something that is also beautiful, whimsical, or emotionally engaging.

Story

When our lives are brimming with information and data, it is not enough to marshal an effective argument. Someone somewhere will inevitably track down a counterpoint to rebut your point. The essence of persuasion, communication, and self-understanding has become the ability also to fashion a compelling narrative.

Symphony

What's in greatest demand today isn't analysis, but synthesis—seeing the big picture and crossing boundaries, being able to combine disparate pieces into an arresting new whole. This is also called systems thinking.

Empathy

The capacity for logical thought is one of the things that makes us human. But in a world of ubiquitous information and advanced analytic tools, logic alone will not do. What will distinguish those who thrive will be their ability to understand what makes their fellow woman or man tick, to forge relationships, and to care for others.

Play

In work and play, there is need for play. The current younger generation has been brought up on computer simulations and learns well in this mode. In addition, there is ample evidence that there is enormous health and professional benefits of laughter, lightheartedness, games, and humor.

Meaning

We live in a world of plenty. We can now pursue more significant desires: purpose, transcendence, and spiritual fulfillment. In addition, with all of the information available, our job now is to make meaning of what is present.

Worksheet #5
Practical Creative Skills

CREATIVE SKILL	CURRENT STRENGTHS	ACTION STEPS
DESIGN		
STORY		
SYMPHONY		
EMPATHY		
PLAY		
MEANING		

PART 2
The Innovative Personality and Skill Sets

❖ Inquisitiveness ❖

🕐 *15 minutes or more*

Purpose

The purpose of this exercise is to build the participants' skills in asking questions and building curiosity.

Materials

✓ An object
✓ A burning sense of curiosity
✓ Worksheet #6

Procedure

1. Divide the larger group into smaller groups of 4 to 7 participants. This can also be an individual exercise.

2. Distribute Worksheet #6 to participants.

3. Give participants a "thing" to examine. This might be an orchid, a pine tree, a potted plant, a car, an air conditioning unit.

4. Each person or group has to develop a list of 25 questions about the item. The questions can be practical or whacky or anything in between. Help the participants by giving them sample questions. For instance, for the pine tree, questions might include:

   a) How many birds have nested in you?

   b) Have you witnessed any murders?

   c) Do they put Christmas lights on you?

   d) What is the most interesting conversation that you have overheard?

   e) How does photosynthesis work?

5. Have the participants return to the larger group and ask each participant to share one or two of their questions.

6. If this is just a warm-up exercise, have the smaller groups go on to the task at hand and develop the 25 questions for their topic.

   a) For instance, this author was once working with a military group trying to reduce the testing time for a hydraulic system for a large piece of equipment. We started by asking 25 questions about the hydraulic system.

   b) For instance, this author was once working with a television station, trying to develop new programming. We started by asking 25 questions about our target demographics (stay-at-home mothers, for instance).

Inquisitiveness (concluded)

Debrief Curiosity is the imperative for innovation. If you wish to be more innovative, start by being inquisitive and asking questions about the world. Recent research suggests that the most innovative executives are the ones who ask the most questions. Great creative geniuses have always asked questions. Leonardo da Vinci found shells at the top of a mountain in Italy and asked why. He also asked why birds could fly and other such "mundane" questions.

Worksheet #6
25 Questions

Build your curiosity by asking questions. Develop 25 questions on a topic: a tree, an on-boarding process, social media, whatever.

1. _____
2. _____
3. _____
4. _____
5. _____
6. _____
7. _____
8. _____
9. _____
10. _____
11. _____
12. _____
13. _____
14. _____
15. _____
16. _____
17. _____
18. _____
19. _____
20. _____

Worksheet #6: 25 Questions (concluded)

21. _____

22. _____

23. _____

24. _____

25. _____

❖ Building Networks ❖

🕐 *60 minutes*

Purpose

The purpose of this exercise is to give participants practice in building networks, which is a key trait of innovative leaders.

Materials

- ✓ Flip chart paper
- ✓ Markers
- ✓ Laptop computers with Internet access
- ✓ A global vision

Procedure

1. Divide the larger group into smaller groups of 3 to 7 participants.

2. The task of each of these groups is to create a "ghost hunting" business. The parents of Baby Boomers are dying off and Baby Boomers desire to communicate with them. You, the would-be proprietor, have always been fascinated by ghosts. You have played around with ghost hunting paraphernalia. You have gone on ghost tours in New Orleans. You have read many books about them and have even thought that you have spotted one or two.

3. If you were going to give this business your best effort, what networking would be useful for you? What organizations would you connect with and/or form partnerships with to build a pattern of success? How might you use social media tools to build this business? Create a media strategy to build your network of success.

4. Allow the groups 30–45 minutes to search the Internet for organizations and people who are engaged in ghost searching, ghost running, ghost exposing, and ghost busting.

5. Have the small groups report-out their media and networking plans to the larger group.

Building Networks (concluded)

Debrief

The groups should have some fun with this exercise. The key learning includes the following:

- Successful innovators know how to network with the right organizations and people.

- They know what the holes in their skill and resource sets are and know that they need partners to fill them.

- Networking is more than a coffee or a lunch; it is a series of mutually beneficial activities and arrangements. For instance, if you create a blog, you might want to mention someone else's similar blog. You might want to mention other organizations of interest to ghost hunters, for example, on your Facebook page.

The key to social media success is getting the buzz about you going in other organizations. It is creating an ecosphere around, for example, ghost hunting with you at the center. This requires intensive networking.

❖ The Rebel ❖

🕐 *60 minutes or more*

Purpose

The purpose of this exercise is to get participants to feel comfortable in challenging the norms and assumptions at their organizations.

Materials

- ✓ Flip chart paper
- ✓ Markers
- ✓ Worksheet #7
- ✓ Invisible sword and shield

Procedure

1. Tell the group that this exercise is often called "assumption testing." Explain that the best and most innovative managers are constantly testing and challenging the norms, assumptions, and status quo.

2. Divide the larger group into smaller groups of 4 to 8 participants.

3. Distribute Worksheet #7 to participants.

4. Ask each group to develop a list of two to three deeply held assumptions.

5. Ask them to challenge those assumptions, one at a time. What if they were not true or necessary? If these assumptions were not true, what could we do differently? What if they could be reversed? What if the opposite were true? Each group should develop a list of implications as if the assumptions were not true. What could be the upside of some of these implications?

6. If the groups are having a hard time thinking about deeply held assumptions, here are some ideas to spur their thinking:

   a) Employees have to be at their desks to be good workers.
   b) Employees have to have suits and ties on to be considered professional.
   c) The only people who want our products are over 50 years old.
   d) We need an internal HR department.
   e) We should not hire people without college degrees.

7. **Variation:** This exercise can be tailored for different groups. For instance, if the group consists of design engineers for autos, some assumptions that may be tested are:

   a) Cars need four wheels
   b) Cars have to cost more than $10,000
   c) Steering wheels need to be round

The Rebel (concluded)

Debrief

It is often difficult to convince employees that they will not be shot if they challenge deeply held assumptions. Just because we challenge the roundness of steering wheels does not mean we are going to immediately change them. Fanciful, bizarre ideas can often lead to productive ones.

It is important to get employees to push the boundaries of traditional thinking. Innovation rarely happens inside narrowly defined lines. Big, game-changing innovations occur by taking big leaps.

Worksheet #7
Assumption Challenging and Testing

"Assumptions are maintained by the hug of history. Yet, history does not guarantee their validity, nor does it ever reassess their validity."

– Michael Michalko

Thus, we must challenge their validity. In this segment, we will challenge the validity of important organizational assumptions.

Blueprint:

1. State a challenge:

    _____
    _____
    _____
    _____
    _____

2. List your assumptions:

    _____
    _____
    _____
    _____
    _____
    _____

3. Challenge your fundamental assumptions:

    _____
    _____
    _____
    _____
    _____
    _____
    _____

Worksheet #7: Assumption Challenging and Testing (concluded)

4. Reverse each assumption. If you can, write down the opposite of each one:

_____
_____
_____
_____
_____
_____
_____
_____
_____

5. Ask yourself, how might I accomplish each reversal?

_____
_____
_____
_____
_____
_____
_____
_____

6. Obtain as many differing viewpoints as you can:

_____
_____
_____
_____
_____
_____
_____
_____

❖ Personality Trait: The Next Box #1 ❖

🕐 *60 minutes*

Purpose

The purpose of this exercise is for participants to see that being innovative often requires "get into the next box" thinking rather than "out of the box" thinking.

Materials

✓ Flip chart paper
✓ Markers
✓ Worksheet #8

Procedure

1. Ask the larger group what they think the phrase "getting outside the box" means. Discuss why that phrase is so popular.

2. Ask what "getting into another box" might mean. How might this process be different? How might one "get into another box"? You might ask, "What other boxes are worth getting into?"

3. You might give examples of organizations that have gone into another box for ideas:

   a) The Bellagio Hotel management went to Italy for ideas on how to innovatively decorate a hotel.

   b) The Rainforest Café management went to a rainforest to build a theme for their chain of restaurants.

   c) The Minnesota Department of Corrections consulted with Target stores to figure out a way to inventory their convicted but released felons.

   d) The Como Park Conservatory personnel went to Japan for ideas on how to design a meditative garden.

   e) Howard Schultz, founder of Starbucks, hung out in Italian coffee bars while developing his idea for Starbucks in the United States.

4. Divide the larger group into smaller groups of 4 to 7 participants. Distribute Worksheet #8 to participants. Ask them to work with one of their organizations or divisions/departments of their organizations to brainstorm what they might be able to learn by going into one of the "boxes" on the worksheet. Or they can think of other boxes (not on the worksheet) to visit and glean ideas from.

5. After 30 to 45 minutes, ask each group to report out an interesting idea or two.

Personality Trait: The Next Box #1 (concluded)

6. **Optional:** Have the participants get back into their small groups. After having heard the ideas from the other groups, ask each participant to claim one high value "box" that they intend to visit in the next six months that might give them valuable ideas.

7. The smaller groups will report out these ideas to the larger group.

Debrief

This is one of the more difficult exercises because it stretches cognitive abilities.

The main point of this exercise is innovation is not just about creating something entirely new. It is about taking someone else's idea and adapting it to your circumstances. You do not need to "reinvent the wheel" in order to be innovative. Look what Schultz of Starbucks did. Coming up with a completely new product often requires engineers, but this kind of "getting into another box" requires just curiosity, acute observation, and seeing with new eyes.

Some of the processes or events on Worksheet #8 will have no relevance for the organization in question. Some of them might have connections. It is important to get the participants to stretch their imaginations in order to make the connections.

Worksheet #8
Other Boxes

- Starbucks coffee delivery system

- UPS package delivery system

- University student registration system

- Target store's return policy system

- Marine boot camp

- Disney World (or another amusement park)

- McDonald's food delivery system

- House-building process

- Art museum preservation, selection, and display

- Minor league baseball entertainment system

- NFL draft process

- Senator or governor selection, campaign, or election process

- Funeral/memorial service, burial process

- Roller coaster operation

- Golf course management

- Cruise ship management

- Scavenger hunt

- Boston Marathon event planning

- Library management

- Preparing a Thanksgiving dinner

- Getting child ready for first day of first grade

❖ Personality Trait: The Next Box #2 ❖

🕐 *60 minutes*

Purpose

The purpose of this exercise is to give participants another method to get into another productive box.

Materials

- ✓ Flip chart paper
- ✓ Markers
- ✓ Laptop computers with Internet access (preferably 2 or 3 per group)

Procedure

1. Ask the larger group what they think the phrase "getting outside the box" means. Discuss why that phrase is so popular.

2. Ask what "getting into another box" might mean. How might this process be different? How might one "get into another box"? You might ask, "What other boxes are worth getting into?"

3. You might give examples of organizations that have gone into another box for ideas.

   a) The Bellagio Hotel management went to Italy to look for ideas on how to innovatively decorate a hotel.

   b) The Rainforest Café management visited a rainforest to build a theme for their chain of restaurants.

   c) The Minnesota Department of Corrections consulted with Target stores to figure out a way to inventory their convicted but released felons.

   d) The Como Park Conservatory personnel visited Japan for ideas on how to design a meditative garden.

   e) Howard Schultz, founder of Starbucks, hung out in Italian coffee bars while developing his idea for Starbucks in the United States.

4. Divide the larger group into smaller groups of 4 to 7 participants. Ask each group to develop a list of five "things" that they would find interesting to understand better. These "things" could be machines or processes or systems. Each group should have access to the Internet. Ask them to go to the website www.howthingswork.com. Call up their list of "things" and study how they work. Take notes on each of them.

5. Ask them to brainstorm what they might be able to learn by going into one of these "boxes." After 30–45 minutes, ask each group to report-out an interesting idea or two.

Personality Trait: The Next Box #2 (concluded)

6. **Optional:** Have the participants get back into their small groups. After having heard the ideas from the other groups, ask each participant to claim one high value "box" that they intend to visit in the next six months that might give them valuable ideas.

7. The smaller groups will report out these ideas to the larger group.

Debrief

The debrief for this exercise is similar to the last one. Once this author was working with a military group and trying to find a faster way to test a hydraulic system. We studied other hydraulic systems in www.howthingswork.com and were able to bring those findings into our problem at hand.

Innovators are curious. They are interested in the way things work. If you are not naturally curious, you probably will not be a natural innovator.

❖ Personality Trait: The Next Box #3 ❖

🕐 *1 – 3 hours*

Purpose

The purpose of this exercise is to help participants naturally get into another box in order to be more innovative.

Materials

- ✓ Flip chart paper
- ✓ Markers
- ✓ Post-it Notes
- ✓ Vivid imagination
- ✓ Worksheet #9

Procedure

1. Divide the larger group into smaller groups of 4 to 7 participants.

2. Ask each individual to focus on a business dilemma/problem/challenge/ situation that warrants some focus. They might want to write these down on Post-it Notes.

3. Distribute Worksheet #9 to participants.

4. In the smaller groups, have one member present their business problem. Have the group choose a "forced associate" from the list on the worksheet.

5. Have members of the group list attributes of that "associate." For instance, if they choose "garden," they might list:

   a) Flowers grow there

   b) Deer eat the broccoli

   c) Basil smells good in autumn

   d) Needs water, mulch, sun, and good soil

   e) Sometimes blighted by beetles and bugs

   f) Some plants might overcome others

   g) Ground might dry up

   h) Attracts bees in August

   i) And so on

Personality Trait: The Next Box #3 (concluded)

6. Have group members connect the problem at hand with the association. For instance, they might say, "Our sales shortfall is like the garden because…" Make as many associations as possible. These connections might bring to light some solutions to the problem. Keep making the connections even if they seem far out or weird or laughable. You never know when a spark lands in the right place.

7. After the groups have tackled a couple of these problems, run a larger group "debrief."

Debrief

This exercise is called forced association because it forces participants to make connections. It is not optional. The connections may seem extreme or silly at first but may yield some valuable insights. These people have presumably tackled these problems for many hours before this exercise. They have probably used linear, left-brain thinking to solve them, without success. This exercise takes them into nonlinear, right-brain, creative thinking that may be more productive, given the chance.

When conducting the debrief, help the group see the connections and honor those connections even though they might not yield a practical solution, *at this point*.

Worksheet #9
Forced Associations

- Garden

- Bar

- Amusement park

- Restaurant

- Professional sports contest (baseball, football)

- Vineyard

- Mountain village

- Shopping mall

❖ Design an Experiment ❖

🕐 *30 – 60 minutes*

Purpose

The purpose of this exercise is to give participants practice in designing experiments.

Materials

✓ Notebook paper
✓ Pens
✓ Flip chart paper
✓ Markers
✓ Laptop computer with Internet access

Procedure

1. Divide the larger group into smaller groups of 4 to 7 participants.

2. Talk with participants about the importance of experimentation in the innovation process. Only when we test our ideas do we get an idea of how good they might be.

3. Have each group choose processes from one of their organizations—a process that might need improvement.

4. Have the smaller groups design experiments around these processes to test new ways of doing them.

5. For each experiment, participants should ask, "What are the unintentional consequences of this procedure?"

6. For each experiment, ask, "What can be learned from this experiment?" "Would the results be the same if we rolled it out to a larger audience?"

7. Have the smaller groups explain their experiments to the larger groups and have all participants ask clarifying questions.

8. Ask participants what it would take to run these experiments.

Debrief

Car companies smash cars into walls to see how crash test dummies survive the impact. Banks set up test branches with innovative layouts to see how customers react. Pharmaceutical companies test out new drugs to see if they work and what side effects may emerge. Then they tweak the ingredients. A *Harvard Business Review* article from the 1980s was entitled "Staple Yourself to an Order." The article suggested that we should all test the procedures that our customers use in doing business with us.

Experiments, and especially failed experiments, can breed innovation. The point is— keep experimenting. You do not know how it will turn out.

❖ Building Innovation DNA into Your Life ❖

🕐 *30 minutes*

Purpose

The purpose of this exercise is to build a plan for individuals to become more innovative in their personal and work lives.

Materials

✓ Worksheet #10
✓ Pens
✓ Imagination
✓ Discipline

Procedure

1. Distribute Worksheet #10 to participants.

2. Re-emphasize to participants how it is important to hold yourself accountable for becoming more innovative.

3. Briefly explain each of the five behaviors associated with Innovation DNA. These five are from the book *The DNA of Innovation:*

   a) **Experimenting:** How to continually conduct experiments in order to improve products and services. Experiments can be simple or complex.

   b) **Questioning:** Get in the habit of asking questions. Like a child, ask why the sky is blue, why fish swim, or why people have two eyes.

   c) **Observing:** In what situations can you be more observant? Observing deeply customers, clients, your products being used, or your competition in action?

   d) **Networking:** You do not have all the keys to the kingdom. Other people, organizations, companies, think tanks, and task forces hold pieces of truth that may be useful to you. Who are these people?

   e) **Associating**: This is not technically a behavior, but a thinking pattern. It is, however, a muscle that can be built. How can you make new connections? Coffee shop and bookstore, car and cup holder, or gourmet food and home delivery.

4. For the first part of this exercise, individuals work alone, quietly trying to fill in Worksheet #10. Ask them to create specific action steps for each of these behaviors. For instance, they may decide to build the habit of asking questions in a specific task force meeting. Or they might go to a specific commercial location and watch their company's product or service being used.

Building Innovation DNA into Your Life (concluded)

5. After about 15 minutes, have participants share their tactics with one or two other people in the larger group. Hearing the action steps of others might spur on more possible action steps of their own.

6. For the larger group, have each individual share one of their action steps.

Debrief

Innovation is not magic or fluff. Sometimes it is just plain hard work. These are behavioral muscles that we—that is, anyone—can flex. It is a matter of practice and being okay with getting it wrong at times. The point is to try it. The point is to observe differently. The point is to ask more questions. The point is to be more curious about one's life.

Worksheet #10
Building Your Innovation DNA

INNOVATION DNA BEHAVIOR	ACTION STEPS
Experimenting	1. _____ 2. _____ 3. _____ 4. _____
Questioning	1. _____ 2. _____ 3. _____ 4. _____
Observing	1. _____ 2. _____ 3. _____ 4. _____
Networking	1. _____ 2. _____ 3. _____ 4. _____
Associating	1. _____ 2. _____ 3. _____ 4. _____

❖ Creative Problem Solving ❖

🕐 *60 minutes*

Purpose

The purpose of this exercise is to give participants practice in using an effective creative problem solving approach.

Materials

✓ Worksheet #11
✓ Worksheet #12

Procedure

1. Distribute Worksheet #11 and Worksheet #12 to participants.

2. Divide the larger group into smaller groups of 3 to 5 participants.

3. Have each participant write down a problem or challenge in his or her life. Using the worksheets, each participant should work through this problem or challenge with the help of their group members. Their group members should coach them through each step using the guidelines from the worksheet. Allow each participant 10–15 minutes.

4. At the end of an hour or so, invite any participants to share the results of the process.

Debrief

This process is used throughout the world. It gives a good framework for working through problems in an organized way.

Worksheet #11
Problem Solving Methodology
(Based on work by The Creative Problem Solving Institute)

- **Objective Finding**
  - *Purpose:* To single out a goal or objective and set its priority
  - *Task:* Develop a list of problems/challenges to work on; settle on the juiciest one
  - *Process question:* What are we really trying to create here?
  - *Outcome:* An aim, general object, or common goal

- **Fact Finding**
  - *Purpose:* To use all of our senses and feelings to deeply examine the objective before selecting important information
  - *Task:* Find out what everyone knows or needs to know about the objective including how they feel about it
  - *Process questions:* Who? What? When? Where? How? What don't we know?
  - *Outcome:* A broad base of information

- **Problem/Challenge Finding**
  - *Purpose:* To seek many new ways to perceive people, situations, and challenges and opportunities from a diverse perspective and to choose the best one
  - *Task:* Generate a variety of problem statements worded for idea stimulation
  - *Process question:* In what ways might we...?
  - *Product:* A single, clear action-oriented problem statement

- **Idea Finding**
  - *Purpose:* To generate alternatives, possible solutions, interesting approaches
  - *Task:* Define and identify the necessary conditions for brainstorming
  - *Process questions:* Substitute? Adapt? Modify?
  - *Product:* An abundance of alternatives, potential solutions, diverse ideas, and creative approaches to problem

- **Solution Finding**
  - *Purpose:* To measure, rank, and examine possible solutions
  - *Task:* Generate a list of criteria for use in evaluating the selected ideas
  - *Process questions:* What needs must be satisfied for the problem to be solved?
  - *Product:* A priority of approaches for solving the problem; choosing a working solution

Worksheet #11: Problem Solving Methodology (concluded)

- **Acceptance Finding**
  - *Purpose:* To identify assistance, objections, acceptance needs, and resources to gain commitment
  - *Task:* Generate all the steps needed to implement the solution
  - *Process questions:* Who? What? Why? When? In what ways might we (IWWMW) gain acceptance?
  - *Product:* A plan of action with times, dates, people, places, and tasks

Worksheet #12
Creative Problem Solving

Objective Finding
Fact Finding
Problem Finding
Idea Finding
Solution Finding
Acceptance Finding

❖ Brainwriting ❖

🕐 *30 – 40 minutes*

Purpose

The purpose of this exercise is to show participants another method for generating a large number of ideas in a short time.

Materials

✓ Blank sheets of paper
✓ Pens

Procedure

1. The room should be set up so that 10 or 12 participants can be seated around a common table. Depending on the size of the group, there might be more than one table.

2. Introduce the concept of brainwriting. Participants probably understand the concept and rules of brainstorming, the process of capturing many ideas out loud in a group setting. Often the most vocal and fast-talking participants get their ideas out there, and quiet, slower but perhaps more thoughtful participants sit quietly. Brainwriting, in contrast, is a silent activity. It allows quiet and more verbal participants to have equal footing in the idea production realm.

3. Each participant should have a blank sheet of paper in front of him or her. Each table should have a problem or a challenge statement. Assign these or have people at the tables decide which issue they wish to work on. Make sure that the tables do not spend too much time on this part of the process.

4. Once the problem/challenge has been decided on, the groups can start the process. Each person should write down one solution on his or her piece of paper and then push the paper to the center of the table. Then they should take a piece of paper that another person has written an idea on and write another idea under theirs. Then they exchange that piece of paper for another one and do it again.

5. The other ideas written on the page should spark other ideas. Some ideas might build on the other ideas; others might emanate from random thoughts. If the sheets fill up, provide another blank one and collect the full one.

6. Call a halt to the activity at any time.

Debrief

Ask for participant reactions to the process. Did it work for them? How did it differ from brainstorming or other processes? Did they feel that they were more fluid with their ideas or less fluid? Was the silence strange? When might they use this process?

PART 3
The Culture of Innovation

❖ Open Source Innovation ❖

🕐 *2 hours*

Purpose

The purpose of this exercise is to give participants an opportunity to craft an open source innovation strategy.

Materials

- ✓ Flip chart paper
- ✓ Markers
- ✓ Laptop computers with Internet access (optional)
- ✓ Not invented here—Not

Procedure

1. Ask participants to define "open source innovation." Simply put, it is the process of obtaining innovative ideas from many sources, including from people outside your organization. You might mention that the packaged goods company Proctor and Gamble made it a corporate goal to raise the number of innovative products *developed on the outside* from about 10% to 50%. A related concept is "crowd sourcing." Crowd sourcing is open sourcing using social network tools.

2. Divide the larger group into smaller groups of 3–7 participants and tell the groups that they will be competing with each other in a contest. Tell the groups that their task is to create a high-powered open source innovation contest for a client. The winner will be the group deemed most likely to acquire the most high-quality ideas.

3. To create these contests, the groups should:

   a) Choose a client organization. It should be an organization with which the group members are familiar: a church, Walmart, Walgreens, a sports team, a food bank, a restaurant, a city, a political party.

   b) Choose an organizational goal or a problem that needs to be solved. Examples might include the following:

   - a church's goal to increase membership
   - a city's plan for a newly acquired vacant lot
   - new product offerings for a company
   - food bank's drive for new volunteers or donations
   - a minor league baseball team's goal to make their games more exciting
   - a college's goal to increase enrollment

4. **Note:** Groups can *either* garner ideas to solve the problem *or* try to solve the problem. It is up to the group.

Open Source Innovation (concluded)

5. Each group needs to distill their idea on to one flip chart page. They will present the idea (in less than five minutes) to the larger group. The members of the larger group can ask clarifying questions.

6. After all groups have presented their ideas, each member of the larger group gets to vote on the plan most likely to succeed. Each participant will place a Post-it Note on the plan that they think is best.

Debrief

Start with a large group discussion on the various open source ideas. What makes for a strong open source campaign? What elements reward participants? What might pique the interest of uninterested people? The larger question is, "How do you enlarge the community that is interested in your endeavors?"

If the group has Internet access, you should suggest that they call up "My Starbucks Idea" (Starbuck's successful open source sight) or General Electric's "Ecomagination Challenge" (General Electric's open source effort at creating energy-saving projects).

❖ Killing the Naysayer ❖

🕐 *30 minutes*

Purpose

The purpose of this exercise is to expose idea-killing phrases and learn how to deal with them.

Materials

- ✓ Flip chart paper
- ✓ Markers
- ✓ Notebook paper
- ✓ Sense of humor
- ✓ Imaginary machete

Procedure

1. Divide the larger group into smaller groups.

2. Ask each group to brainstorm phrases that kill ideas that they hear in their organization. Examples might include:

   a) That would not work here.

   b) We tried that five years ago.

   c) Top management will never go for that.

   d) It is not our job to come up with new products.

   e) We can never afford something like that.

3. After 5 to 10 minutes, have each group present some of their phrases to the larger group.

4. Have the smaller groups reconvene. Ask them to develop one or two "comebacks" for each of the killer phrases. These should be appropriate and effective comebacks, not just insults or put-downs.

5. Have each group present the "comebacks" to the larger group.

Debrief

This exercise can lead to the larger topics of organizational impediments to change and innovation. It may lead to underlying cultural constraints, some of which cannot be removed by the participants. You might want to develop two lists: "cultural constraints we can move" and "cultural constraints that are immutable."

You should not allow this conversation fall into a "complaint" session where the participants feel that the organization is too powerful and entrenched in order to change. You should bring the conversation back to what is possible within the organizational constraints.

❖ Where do babies (innovative ideas) come from? ❖

🕐 *2 hours (to weeks and months—this could be part of a strategic planning process)*

Purpose

The purpose of this exercise is to push the participants to think about new sources of innovative ideas. Participants will also develop an action plan for capturing those ideas.

Materials

✓ Flip chart paper
✓ Markers
✓ Diligence and persistence
✓ Worksheet #13
✓ Laptop computers with Internet access (optional)

Procedure

1. Distribute Worksheet #13 to participants.

2. As a large group, discuss each of these items and develop a list of examples for each of them. For example:

   a) *Unexpected occurrences:* 9/11, tsunami, oil spill, earthquake

   b) *Incongruities:* high cost of AIDS drugs

   c) *Process needs:* faster, fewer steps, digitalization

   d) *Market and industry changes:* outsourcing, price compression, savvy consumers

   e) *Demographic changes:* aging Baby Boomers, Tech-savvy Millennials

   f) *Changes in perception:* green is gold, thrifty is good

   g) *New inventions:* digital technology, new medicines, electric engines

3. Divide the larger group into smaller groups. Have each small group choose one or two of these "sources of innovation." The groups need to develop an organizational action plan around one or two of these sources. For instance, if they choose "changing demographics," they can choose a subgroup or two and create a mini-marketing plan around a product or service to meet the needs of this group. These product/service offerings should include the following:

   a) product/service description

   b) promotional strategy, including social media strategy

   c) pricing strategy

   d) distribution strategy

   e) customer service strategy

4. Have the groups present their product/service ideas to the larger group.

Where do babies (innovative ideas) come from? (concluded)

Debrief

The point of this exercise is to have participants expand their thinking about the origins of innovations. Ask participants about their process in doing this exercise. The responses might range from frustration to fun. Yes, it is frustrating trying to create an entire product offering in a short time, but you might reinforce that this is just practice in order to get them in the habit of finding innovations.

Worksheet #13
Sources of Innovation
(Adapted from *Innovation and Entrepreneurship* by Peter Drucker)

- Unexpected occurrences

- Incongruities

- Process needs

- Market and industry changes

- Demographic changes

- Changes in perception

- New inventions

❖ Failure Notebook ❖

🕐 *30 minutes*

Purpose

The purpose of this exercise is to begin to create a productive failure/learning book for an organization.

Materials

✓ Ability to laugh at oneself
✓ Laptop computer for each small group

Procedure

1. Tell the larger group that it is time to laugh at ourselves and have fun with it.

2. Divide the larger group into smaller groups of 3 to 7 participants.

3. Have each group brainstorm mistakes they have made and what they have learned from those mistakes. Limit the brainstorm, if you can, to work-related issues. Others can then understand the contexts of these mistakes.

4. Have a scribe for each group type these in a database or on a PowerPoint.

5. After about 15 minutes, have each group report-out their findings.

6. To make it more fun or funny, take a vote on which was the biggest blunder.

Debrief

Tell the larger group how some organizations institutionalize their failures. Have the participants talk about their experiences in this exercise. Did they feel shame or humiliation? Or did it feel good knowing that others made serious errors?

Make the point that if we can learn from our mistakes, we can learn together. If we shared like this often, we would not make the same errors over and over again.

❖ Faces of Innovation ❖

🕐 *30 minutes*

Purpose

The purpose of this exercise is to help participants understand that there are different ways of being innovative.

Materials

- ✓ Worksheet #14
- ✓ Notebook paper
- ✓ Pen
- ✓ Broad working definition of *innovation*

Procedure

1. Divide the larger group into smaller groups of 3 to 7 participants.

2. Distribute Worksheet #14 to participants and allow a few minutes for them to read it.

3. Explain to participants that innovation involves many roles and many hats.

4. Ask them to think about which of these ten faces that they play out in the workplace. Have them present these to their small groups. Have them explain how they play these out and which behaviors they exhibit.

5. After all the groups are finished with that part, ask them to think deeply about which roles are missing in their organization. How could they fill those roles? What kinds of people could fill those roles? How could a group, using a disciplined approach, fill those roles?

6. Facilitate a large-group discussion to finish the exercise.

Debrief

Diversity is important to the innovation process. A room full of wild-eyed, creative thinkers may not accomplish much. In organizations, different skill sets are necessary to be innovative. In many cases, no role is any more or less important than another role. Discuss each of these faces and how participants might be able to uncover and expand their own and others' innovative techniques within an organization.

Worksheet #14
The Ten Faces of Innovation
(Adapted from *The Ten Faces of Innovation* by Tom Kelley)

The Learning Personas

Anthropologist: observes human beings and develops a deep understanding of how people interact physically and emotionally with products, services, and spaces

Experimenter: prototypes new ideas continually, learning by an enlightened trial and error

Cross-Pollinator: explores other industries and cultures and then translates those findings and revelations to fit the unique needs of your enterprise

The Organizing Personas

Hurdler: knows the path to innovation is strewn with obstacles and develops a knack for overcoming or outsmarting those roadblocks

Collaborator: helps bring eclectic groups together and often leads from the middle of the pack to create new combinations and multidisciplinary solutions

Director: not only gathers together a talented cast and crew but also helps spark their creative talents

The Building Personas

Experienced Architect: designs compelling experiences that go beyond mere functionality to connect at a deeper level with customers' latent or expressed needs

Set Designer: creates a stage on which innovation team members can do their best work, transforming physical environments into powerful tools to influence behavior and attitude

Caregiver: builds on the metaphor of a health care professional to deliver customer care in a manner that goes beyond mere service

Storyteller: builds both internal morale and external awareness through compelling narratives that communicate a fundamental human value or reinforce a specific cultural trait

❖ Event/Pattern/Structure ❖

🕐 *60 minutes*

Purpose

The purpose of this exercise is to introduce systems thinking as an important part of innovation.

Materials

✓ Flip chart paper
✓ Markers
✓ Wide-angle view of life

Procedure

1. Divide the larger group into smaller groups of 3 to 7 participants.

2. Introduce the concept of event/pattern/structure:

   a) *Event:* actual event occurrence

   b) *Pattern:* series of like events or occurrences

   c) *Structure:* the underlying structure that is creating the pattern

3. Example:

   a) *Event:* a fender-bender car accident at a particular corner

   b) *Pattern:* series of the events or occurrences

   c) *Structure:* the underlying structure that is creating the pattern

4. Ask each group to choose a negative event or occurrence in their workplace. Have them figure out if there is a pattern. (Examples might include chronically late reports, dirty restrooms, or delayed shipments.)

5. Have the groups brainstorm and figure out the underlying structures that are causing these patterns.

6. Have the groups brainstorm (they can use one of the techniques delineated in this book) possible "fixes" for these underlying structural problems.

7. Have the groups report-out their findings to the larger group.

Debrief

Systems thinkers think in terms of underlying structures and not just events. It is important to look deeply for the underlying problems, not just surface issues. Otherwise, you might be solving the wrong problem, and the solution will not last.

This author once worked with a second-grade teacher. When a particular little boy acted out inappropriately, she would take away his recess time. Guess what? That just exacerbated the problem. Why? Because he was severely ADHD, taking recess away just made him more fidgety and prone to acting out. A better solution in this case might have been to have him sweep the floor or run files from room to room.

❖ Trend Spotting ❖

🕑 *1 – 2 hours*

Purpose

The purpose of this exercise is to give participants practice in spotting trends and translating those trends into possible innovations.

Materials

✓ Notebook paper
✓ Pens
✓ Fiery imagination
✓ Laptop computers or iPads (one for every two participants)

Procedure

1. Divide the larger group into smaller groups of 4 to 8 participants. Distribute the laptops/iPads.

2. Explain that the purpose of this exercise is to give participants practice in spotting and working with trends in the marketplace.

3. Have each group choose an industry: auto, coffee shop, oil, fashion, craft shops.

4. The first task of the groups is to investigate the industry. They can use Google or another Internet search engine. They could type in "office furniture, innovations."

5. While half the group "surfs," the other half should take notes. What trends, innovations, initiatives, new policies, or events are happening in that industry?

6. After about 20–30 minutes, the groups should switch to an innovation mode. Given these trends and other industry news, what innovations is this industry ready for? What needs are being created by the trends, innovations, and events? Have the groups spend 20 minutes brainstorming possible innovations.

7. Have the smaller groups report-out to the larger group.

Debrief

Innovations spawn other innovations. Initiatives spawn innovations. Events create space for innovations. For instance, cell phones created a billion-dollar ringtone industry. Gas-fueled autos spawned a huge oil-drilling industry.

To get ahead of the curve, individuals and organizations must be constantly monitoring trends and innovations. What is next in that business or industry? You can ask how organizations and individuals can methodically and systematically collect trends and act on them. This will be a rich conversation.

❖ Creative Collaboration ❖

🕐 *1 hour*

Purpose

The purpose of this exercise is to help participants use concepts of creative collaboration and evaluate their organizations with them.

Materials

✓ Flip chart paper
✓ Markers
✓ Notebook paper
✓ Pens
✓ Worksheet #15

Procedure

1. Divide the larger group into smaller groups of 4 to 8 participants.

2. Distribute Worksheet #15, which lists principles of creative collaboration to participants.

3. Have each group choose an industry: auto, coffee shop, oil, fashion, craft shops.

4. Discuss the concept of creative collaboration with participants. Creative collaboration is a stepping stone to innovation. The more creative collaboration in a group, the more likely the group is to produce new, productive ideas.

5. Each group should divide a flip chart page in half vertically. They should label each column "Helpful" and "Not Helpful" (to Creative Collaboration). Using the worksheet ideas, they should brainstorm behaviors and attitudes for these two columns.

Debrief

Very little is done in isolation. To tackle complex problems, we clearly need creative collaboration at its best. Everyone in organizations needs to learn how to play well in the sand box.

During this debrief, if a participant says something like "communicate better," ask him or her to go deeper. What exactly do they mean by communication? In what context? Ask participants for specific action steps to make their workplace more conducive to creative collaboration.

Worksheet #15
Principles of Creative Collaboration

Open, advanced communication: Participants should be trained on how to build on to each other's voices and incorporate respectful give-and-take as part of the creative process.

Diversity of voices: Many different voices help with creative collaboration. People with diverse perspectives see things in differing but interesting and complex ways.

Community space for communication: Nurses use logs. Many new computer programs allow for joint creation of documents or drawings. Bulletin boards allow for messaging during non-meeting times.

Watering holes: Creative collaboration necessitates a place where people can come together naturally, usually around food or coffee. It may be a coffee shop, a table out in the open, or a conference room.

Open time: There needs to be time dedicated to innovation and collaboration. It cannot be an afterthought.

"Yes, and" rather than "yes, but": This is a subtle but powerful distinction. It is an active push to get negative language away from the act of creation.

Rapid prototyping: It does not matter if it is a three-dimensional representation or a sketch on a pad of paper. It is important to create a working prototype in order to have something to which to react.

❖ Building an Innovative Culture I ❖

🕐 *30 minutes*

Purpose

The purpose of this exercise is to give participants a start on building an innovative culture in their workplace.

Materials

✓ Worksheet #16
✓ Pens
✓ Laptop computer with Internet access (optional)

Procedure

1. Divide the larger group into smaller groups of 4 to 7 participants.

2. Distribute Worksheet #16 to participants.

3. Have the groups fill in the worksheets as well as they can. Have them be creative. The goal is to have an action plan for creating an innovative workplace. They might get into a mode of "we cannot do this because…" but help them past this phase. There will always be resistance to new ideas.

4. Have participants use the computers to look up books or articles that might be useful.

5. Have participants share their ideas with others in the larger group.

Debrief

Innovation can be fun. Make it so. Build innovation activities into the calendar; do not wait for them to happen on their own. Innovation activities can be high energy. Add food and playfulness to the mix.

Emphasize that innovation is doable. It does not necessarily take a huge budget, creative geniuses, or expensive retreats to a mountaintop hideout, even though those are nice. Departments can take their own innovation initiatives and make them work. These efforts can become infectious to those around the initiators.

The important part is to start somewhere, even if it is just a once-a-month, lunch book club.

Worksheet #16
Building an Innovative Culture

INITIATIVE	ACTION #1	ACTION #2	ACTION #3
Innovation Book Club			
Failure Party			
Innovation Story Communication			
Prototype Development Party			
Innovation Room Development			
Trend Spotting Group			
Customer Probing Circle			
Process Redesign Initiative			
Other			

❖ Building an Innovative Culture II ❖

🕑 *1 hour*

Purpose

The purpose of this exercise is to help an organization gauge its openness to innovation.

Materials

✓ Critical eye
✓ Worksheet #17
✓ Pens

Procedure

1. Divide the larger group into smaller groups of 4 to 7 participants.

2. Distribute Worksheet #17 to participants.

3. Ask participants to look critically at their organization and the checklist. In each category, how does the organization do well, and when does it fall short?

4. Have smaller groups report-out their findings to the larger group.

Debrief

What makes for an innovative workplace? Participants can give many deterrents to innovation, but can they conceive and execute a culture that encourages innovation? This exercise allows them to break down parts of a culture and examine it, and to see how to improve it. The biggest takeaway from this exercise can be the conversations that take place.

Ask participants which elements are missing and how to bring those forward. Ask them which of those elements on Worksheet #17 are hardest to build into the culture.

Worksheet #17
A Culture of Innovation

Write down two or three examples of each of these elements that are embedded in your culture.

- **Open Dialogue**

    - _____
    - _____
    - _____

- **Encourage New Ideas**

    - _____
    - _____
    - _____

- **Sufficient Resources**

    - _____
    - _____
    - _____

- **Reinforcement**

    - _____
    - _____
    - _____

- **Respect**

    - _____
    - _____
    - _____

- **Opportunity**

    - _____
    - _____
    - _____

Worksheet #17: A Culture of Innovation (concluded)

- **Long-Term Perspective**

  - _____
  - _____
  - _____

- **People Advantage**

  - _____
  - _____
  - _____

❖ Scaffolding: Toward a Culture of Innovation ❖

🕐 *30 – 60 minutes*

Purpose

The purpose of this exercise is to evaluate the personnel in an organization for readiness for innovation.

Materials

✓ Worksheet #18
✓ Pens
✓ Critical, self-reflective eye
✓ Laptop computers with Internet access

Procedure

1. Divide the larger group into smaller groups of 4 to 8 participants.

2. Distribute Worksheet #18 to participants.

3. Facilitate a discussion of each of the five scaffolds for effective innovation.

   a) **Emotional intelligence.** New innovations require intense collaboration of units, subunits, people, and departments. Innovators have to be able to get along with each other. Emotional intelligence consists of several different facets:

   - Self-awareness
   - Self-management
   - Self-motivation
   - Empathy
   - Social skills

   b) **Innovation skills.** That is what this entire book is about—building different innovation muscles.

   c) **Process training.** Some of this training includes "quality" or "project management" tools: flow charting, value stream mapping, Gantt charts, and the like. The five-step innovation process must also be learned and practiced by employees.

   d) **Motivation.** Sometimes, "we need new products to stay in business" or "we need to speed up our processes to stay competitive" is not enough to motivate employees. Sometimes, contests, prizes, awards, or other kinds of recognitions are needed. Sometimes, in the short run, money is a good incentive.

   e) **Systems thinking.** Employees often need a primer on systems thinking: systems archetypes, systems principles, pattern recognition, and systems failures. Systems thinking is a way of looking at the world, a way that is scaffolding for many disciplines.

Scaffolding: Toward a Culture of Innovation (concluded)

Debrief

Scaffolding is an educational concept. Students need to understand Algebra before they move on to Calculus. Students need to complete Spanish I before they move on to Spanish II. The same runs true for innovation. We need to master certain concepts before we can fully engage in others. Yes, there are plenty of examples of lone, eccentric geniuses who rant and rave at others and then invent great projects. Those examples are becoming less and less prevalent. More and more, close creative collaboration creates great innovations. Before diving into innovation projects headlong, organizations should bring their employees up to a certain level of competence in these skills.

Worksheet #18
Scaffolding for Innovation

SCAFFOLD	CURRENT STATE	ACTION STEPS
EMOTIONAL INTELLIGENCE		
INNOVATION TRAINING		
PROCESS TRAINING		
MOTIVATION		
SYSTEMS THINKING		

PART 4
The Innovative Process

1. Probing the constituency
2. Observe the real situation
3. Develop new concepts
4. Converge and build prototypes
5. Implementation

❖ Open Source Innovation ❖

🕐 *1 – 4 hours*

Purpose

The purpose of this exercise is to help participants become better observers. People who observe more acutely are more likely to be better innovators.

Materials

✓ Notebooks
✓ Pens
✓ Digital cameras
✓ Computer projector
✓ Laptop computers
✓ Eagle eyes

Procedure

1. This exercise can be done with groups or with individuals.

2. If in groups, divide the larger group into teams. Provide a digital camera to each team.

3. Have each team choose a location to study where consumers consume "something." Send teams out to these locations to observe consumers. Have them take copious pictures of consumers engaged in their tasks. Ask them to observe for an hour or so. Ask them to look for things such as

   a) consumer behaviors;

   b) the way consumers do the work they do;

   c) hassles consumers seem to be having;

   d) jobs that need to be done;

   e) the context of the consumption;

   f) anything else that is interesting.

4. When the teams come back to the seminar location, have them create a PowerPoint from their pictures and notes. They should be ready to tell a story to the larger group about what they saw. Make sure they differentiate between their observations and their inference. (For example, "a person walking into a restaurant" is an observation. "Because they are hungry" is an inference.)

5. Some groups might be confused as to an appropriate place to observe. Examples of places and consumers to observe might be

   a) buyers loading children and groceries into cars in a grocery store parking lot. Maybe the grocery store company wants to make this process easier and safer.

Open Source Innovation (concluded)

b) people at a baseball game (if you have the time). Maybe the owner of the ball team wants to make the experience a more memorable one.

c) people at a coffee shop—what do they use it for? How can the coffee shop raise the level of that experience?

d) people crossing the streets at a busy urban intersection. How might the city's highway department make the experience safer?

6. Have each team present their insights to the larger group.

Debrief

This can be a powerful and fun exercise. Most participants have been at these places of observation, but have not observed deeply what is happening. They will see things that they have never seen before. They will have new insights. These insights may lead to innovations.

The purpose of this exercise is to promote deep thinking. Make sure that participants can discern the difference between just "casually seeing" and "deeply observing."

Early (1920s) car salesmen observed farmers buying Model T Fords and then immediately tearing the back seats out. This observation led to the creation of a pickup truck. A person observing car drivers holding cups of coffee between their legs created cup holders in cars. (Yes, some of you readers are too young to remember cars without cup holders. But many of us do remember.)

❖ Deep Inquiry ❖

🕐 *60 minutes*

Purpose

The purpose of this exercise is to give participants practice in deep inquiry toward the goal of innovation.

Materials

- ✓ Notebook paper
- ✓ Pens
- ✓ Desire for deep understanding

Procedure

1. This exercise can be done in groups of 2, 3, or 4. Divide the group accordingly.

2. Ask one participant from each group to think of an activity that they love to do.

3. Ask the other members of each group to begin to ask the designated person about his or her hobby or activity. The questions should become progressively deeper. Some questions might be:

   a) How do you feel when you are doing this activity?

   b) This activity is like a _____ when I am engaged in it.

   c) If this activity were an animal, it would be _____.

   d) What do you wish for when you are doing this activity?

   e) Who do you wish to share this activity with?

   f) What is one thing that would make this activity even better?

   g) What do you think about when you are doing this activity?

   h) What deep needs is this activity filling?

4. After asking these questions for about 20 minutes, stop. Have the small group review the data, and think of any products or services that would make this activity better.

5. Go on to the next person and do the same kind of questioning. Do this for the time length that you have.

Debrief

This exercise helps build the muscle of questioning. Trying to understand deep needs, aspirations, hopes, and dreams can help the process of innovation. The better one understands the psyche, the better one can help fulfill the needs of that psyche through products and services.

❖ Creative Collage Making ❖

⏱ *2 – 4 hours*

Purpose

The purpose of this exercise is to show how visual stimuli can be used to stimulate innovation.

Materials

✓ Old magazines full of pictures
✓ Glue sticks
✓ Poster boards
✓ Markers
✓ Scissors
✓ Post-it Notes
✓ Playfulness
✓ Digital cameras (optional)

Procedure

1. Divide the larger group into smaller groups of 4 to 7 participants and let them know that they are going to have fun today.

2. Ask each group to choose a psychographic profile to dive into, for instance:

   a) Suzy, 25-year-old, single, upscale urban professional

   b) George, 67-year-old, retired, living on fixed income with wife in first ring suburb

   c) Joan, 36-year-old, financially struggling, single mom

   d) Sam, 23-year-old, urban graduate student with high aspirations

3. **Phase 1:** Have each group page through the magazines and find pictures of people, products, retail locations, and other artifacts that fit this personage. Have them cut out these pictures and create a collage that tells the story about this psychographic. They might want to write down notes about a day in the life of this individual.

4. Have a representative from each group report out to the larger group. These reports should include things such as

   a) likes and dislikes;

   b) media used;

   c) retail establishments frequented;

   d) dreams and aspirations;

   e) political outlooks.

   Note: This phase might take some imagination. If you were real marketing professionals, you would have hard data and elaborate reports from which to work.

Creative Collage Making (concluded)

5. After each presentation, have all of the other participants write down one "insightful, brilliant, and probing" question on a Post-it Note. Collect the questions.

6. At the end of the presentations, distribute the Post-it questions to the appropriate groups. The groups discuss the questions among themselves.

7. **Phase 2:** Back in the smaller groups, each group will develop one to three new product/service offerings for their demographic target. They are to elaborate on these product/service offerings and include

   a) price points;

   b) distribution channels;

   c) marketing strategy;

   d) promotional strategy.

8. Have a member of the small groups report-out to the larger group. Optional: You can have other participants ask elaborating questions to the presenting group.

9. **Variation:** Instead of using magazines, you can send the groups, with digital cameras, out into the world. They can visit stores, restaurants, bus stops, and places like that and take pictures of people who seem to fit their demographic profile. This variation will only work if you are conducting the workshop in an area with many people. Instead of creating a collage, the groups can create a slide show on that personality profile.

Debrief

Debrief questions might include:

- Did you feel like a marketing guru?
- How well did you get into the heads of your psychographic profile?
- Did the visual prompts help your creative process?
- What additional questions would you have for your "Suzy" or "George"?
- What other methods would you employ to get into their heads?

❖ Disassembling ❖

🕐 *1 – 2 hours*

Purpose

The purpose of this exercise is to build the curiosity of participants.

Materials

- ✓ Flip chart paper (optional)
- ✓ Markers
- ✓ Notebook paper
- ✓ Inquisitive nature
- ✓ Laptop computers with Internet access
- ✓ Old and broken appliances and machines (for example, computers, blenders, light switches, printers, de-humidifiers, microwaves, TVs)
- ✓ Screwdrivers
- ✓ Hammers
- ✓ Pliers
- ✓ Wrench set

Procedure

1. Divide the larger group into smaller groups of 3 to 4 participants.

2. Give each group one or two of the appliances and some tools. Tell them to disassemble the appliances. Their goal is to totally understand how the appliance works. They can use the Internet and the website www.howthingswork.com to help them with this.

3. After they have disassembled and understand their appliances, have a member from each small group explain and describe it to the other members of the larger group.

4. Have the small groups then get back together and study the appliance. What innovation or innovations would make the appliance more effective? What other bells and whistles could be added to make it sexier or better?

5. Have a member from each small group report-out these innovations to the larger group.

6. **Optional:** Have the small groups put the appliances back together and workable. (This might be really fun.)

Disassembling (concluded)

Debrief

There is no "wrong" way to do this exercise. Participants should have fun. You might ask: What does this have to do with your job? Hopefully they talk about the curiosity to take apart existing processes and ways to do things and look at them with new eyes. One way to innovation is to take apart and totally understand what you currently have in order to improve it or build on it. This is a discipline not solely in the purview of engineers. Innovation is everyone's job.

You may want to mention that if they do this at home to please unplug the appliances before disassembling.

❖ Question Circle ❖

🕐 *45 minutes*

Purpose

The purpose of this exercise is to allow participants to formulate deep questions. This exercise also allows participants to sit with unanswered questions to their own concerns.

Materials

- ✓ Post-it Notes for everyone
- ✓ Inquisitiveness
- ✓ Laptop computer (optional)

Procedure

1. Rearrange the room so that groups of 10 to 12 participants can sit in circles.

2. Ask each participant to think of a difficult business problem/challenge/dilemma that they are facing. Optional: Have them write down the problem. Allow 3 to 5 minutes of thoughtful silence.

3. Each participant will have the opportunity to present his/her question to the group in a circle. As each person presents her/his problem/challenge/dilemma, allow the group several minutes of thoughtful reflection. Members of the group can ask clarifying questions to the presenter.

4. Each participant should develop at least one insightful question about the presented problem/challenge/dilemma. They can write them down on Post-it Notes if desired.

5. Then, going around the circle, each participant presents their first question. The problem presenter **says nothing.** The presenter just listens and thinks about what she/he is hearing.

6. After one round of the circle, the floor is open for anyone to ask additional questions. This continues until there is silence. The presenter thanks the group, and the group moves on to the next presenter.

7. A scribe could be capturing all of these questions on the laptop to distribute to all of the presenters. Or the presenter could collect the Post-it Notes as a remembering device.

8. Once all of the participants have presented their problems/challenges/dilemmas, have a large group discussion about the process.

Question Circle (concluded)

Debrief   Ask the larger group how it was to be part of the process. Most presenters probably had a hard time being quiet and not trying to answer the questions. You might mention that often silence breeds thoughtfulness. Also, too often we think that we have the answers when we really do not.

The points of this exercise are to

- learn how to formulate powerful questions;

- build a learning community;

- use astute questioning as part of the innovation process;

- learn to sit with unanswered questions rather than jumping to the punch;

- learn how to be helpful to colleagues.

❖ Deep Empathy ❖

🕐 *2 hours*

Purpose

The purpose of this exercise is to allow participants to be better anthropologists, exploring an important issue for their human friends.

Materials

✓ Paper and pens (or laptop computers)
✓ An insatiable case of curiosity
✓ Worksheet #19
✓ Flip chart paper and markers (optional)

Procedure

1. There is not an optimum size for this exercise. Each group should have two sub-groups of at least four participants. There can be many of these groups.

2. Have each group choose who will be the questioners and who will be the subjects. These groups will change roles at each round.

3. Distribute Worksheet #19 to participants.

4. Each group must start with one event or experience. Examples might include the following:

   a) going to college

   b) having a physical at the doctor's office

   c) going to a professional sports event

   d) going to a parent/teacher meeting at school

   e) shopping for intimate apparel

   f) going to a movie

   g) cooking a three-course meal

   h) attending a wake or funeral

   i) burying a pet

5. The questioners will attempt to understand fully and deeply the participants' needs, desires, wishes, and feelings about taking part in this event or experience.

6. Allow the group 30 minutes. The questioners should jot down notes. At the end of the allotted time, ask each group (two subgroups) to brainstorm new services or products that may enhance this event or experience. Have one group member take notes on this.

Deep Empathy (concluded)

7. **Optional:** The questioners may want to organize their data in the following way. Create a 4-quadrant grid on a flip chart page. Label the quadrants: "say," "do," "think," and "feel" and put remarks into one of these categories.

8. Ask the subgroups to switch roles and choose another event or experience. Again, allow the questioning to go on for 30 minutes.

9. At the end of the 30 minutes, allow the group to brainstorm for new product/ service ideas.

Debrief   Ask the participants to reflect on the exercise:

- Was it fun?

- Was it uplifting?

- Did it get them to think in different ways?

- Can they see an application for their organization?

- Did it help them build their curiosity levels?

- Do they think that they dug deeply enough?

- How could they have dug more deeply?

Your final comment might be "Innovators must love people and be curious about them."

Worksheet #19
Probing Questions

1. Describe your highest moment at the event or experience (EE).

2. Describe your lowest moment.

3. Tell a funny story about attending the EE.

4. Were you ever embarrassed at the EE?

5. Did you ever feel sad at the EE? What might have taken that sadness away?

6. When have you known of a friend to have a bad time at the EE? What made it bad/sad/unhappy/ugly?

7. If the EE were an animal, what kind of animal would it be? Why?

8. If you were to redesign the space where the EE happened, how would you do so?

9. If there were a new rule concerning the EE, what would it be?

10. If the EE were a color, what color would it be?

❖ Photo Wall ❖

🕐 *1 – 4 hours*

Purpose

The purpose of this exercise is to show how visuals can be used in the innovation process.

Materials

- ✓ Flip chart paper
- ✓ Markers
- ✓ Tape
- ✓ Wall space
- ✓ Digital cameras (one per person, preferably)
- ✓ Photo printer
- ✓ Photo paper

Procedure

1. This activity can be done in either a small or a large group. Each group can choose a different activity or use the same one.

2. Send the group (or groups) out to observe an activity. Possible activities include the following:

   a) grocery shopping

   b) clothes shopping

   c) crossing urban streets

   d) using mass transportation

   e) going to the emergency room at a local hospital

   f) going to a beach or swimming pool

   g) going to a local pub or sports bar

   h) shopping at a mall

3. Have the groups take as many photos as possible of the activity. Have them print the photos, either with a photo printer or at a nearby drugstore or big box store. Have the groups return to the workshop room after a specified time.

4. Have the groups tape their photos on a wall and look at them. If they want, they can organize them how they choose.

5. Have each group label three flip chart pages "what," "how," and "why."

6. Have them fill these flip chart pages from what they see in these photos.

Photo Wall (concluded)

7. Using their photos, their organizational scheme (if they have one), and their flip chart pages, have the groups brainstorm possible innovations for the activity they have taken photos of. For instance, if they have photo-documented a pub, what services or products might make this a richer experience for participants?

8. Have the smaller groups report-out to the larger group.

Debrief   Ask participants how the experience went for them. Ask them if they saw things in the photos that they did not see with their own eyes. Discuss how visual images can spark ideas that words or conversation may not be able to.

❖ Can you hear me now? ❖

🕐 *60 minutes*

Purpose

The purpose of this exercise is to show how an organized brainstorming session can lead to actions taken and problems solved.

Materials

✓ Flip chart paper
✓ Markers
✓ Post-it Notes
✓ Pens
✓ Worksheet #20

Procedure

1. Distribute Worksheet #20 to participants.

2. Ask participants to name some of their organizational problems.

3. One of the issues that will always come up is communication. Typically, the conversation stops there. But what do they mean by poor communication? This exercise gives an opportunity to dig deeper into this issue.

4. Ask participants to write on Post-It Notes 5–10 specific communication problems in their organization and to stick the Post-it Notes on a wall or on a table.

5. Ask the group to organize the Post-it Notes into themes. This might be chaotic with a larger group, so if it is large, you may want to divide the group into two smaller groups.

6. To do this, you might have them create a flip chart page for each theme and put the appropriate Post-it Notes on the correct flip chart page.

7. When that is accomplished, have each smaller group choose one of the themes (flip chart page with Post-it Notes).

8. Have each of the smaller groups put together a series of action items in order to alleviate the problem. Each action item needs to have a champion, a team, a timeline, and an assessment tactic.

9. Have each of the smaller groups report-out their findings and action steps to the larger group.

Can you hear me now? (concluded)

Debrief

Some words are becoming cliché-ish in western culture. "Communication" is one of them. This exercise attempts to add meaning to that word. During the debriefing, ask participants how energized they feel. Ask them how doable their action plans seem.

You may want to talk about the entire process of communication. Why is it that we have more communication tools than ever and yet communication often seems worse than ever?

Ask participants what other organizational problems might be able to be broken into bite-sized pieces and solved.

Worksheet #20
Brainstorming Rules

- Encourage weird ideas

- Focus and refocus on issue

- One conversation at a time

- Build on each other's ideas

- No negative feedback

❖ Redesign Rooms ❖

🕐 *1 – 4 hours*

Purpose

The purpose of this exercise is twofold: to create an innovative space for learning and to give participants a chance to use the process of innovation.

Materials

- ✓ Flip chart paper
- ✓ Markers
- ✓ Notebook paper
- ✓ Colored pencils
- ✓ Rulers
- ✓ Other drawing material, as needed
- ✓ A laptop computer or similar device or two with Internet access for each group

Procedure

1. Divide the larger group into smaller groups of 4 to 7 participants.

2. Tell the groups that their objective is to design a room for maximum learning. Here are the rules:

   a) Money or space is not an issue.

   b) Use any technology that is available.

   c) Use any furniture available—use the Internet to search for furniture.

   d) Use any resources at hand for ideas.

3. Allow the groups ample time to research and brainstorm and develop their ideas.

4. Have participants report-out their final prototype to the larger group. They can use a PowerPoint slide or flip chart pages.

Debrief

After the report-outs, ask the groups to talk about the steps in their process.

- What were the steps they used?
- Which steps seemed to energize? Which steps seemed to deflate?
- Where did they get stuck?
- Did they converge and diverge at times?
- How could they have improved their process?

❖ Two Circle Critique ❖

🕐 *1 – 2 hours*

Purpose

The purpose of this exercise is to demonstrate a critique process that could lead to an innovation.

Materials

- ✓ Flip chart paper
- ✓ Markers
- ✓ A product that many people use
- ✓ A probing sense of curiosity

Procedure

1. Choose a product or service that participants use, have used, or are familiar with. It could be a CD player, the conference room, a blender, a local coffee shop, a flip chart pad, a barber shop. Conversely, you could use a hypothetical product that has not been introduced into the marketplace. It could be in proto-type form or sketched out on paper.

2. Divide the larger group into two smaller groups. If the group is too large, create four smaller groups.

3. Have one group create and sit in an inner circle of chairs and the other group should sit in chairs around them.

4. The outer group should have two members standing at flip charts. The flip charts should be divided into four quadrants: green light or positives, red lights or criticisms, mild or wild ideas, and puzzling questions.

5. The outer group should question the inner group about the experience, service, or product. They should probe as deeply as possible. As the inner group responds, the scribes should put the responses in the appropriate boxes.

6. After 20–30 minutes, the entire group should discuss the results. Brainstorm an open discussion about possible innovations based on the grid. How can the product/service/experience be improved?

7. Have the groups reverse the seating so that the inner circle becomes the outer circle and vice versa. Redo the process with another product/service/experience.

Two Circle Critique (concluded)

Debrief

This exercise helps participants build on whatever exists already. Ask the following questions in the debriefing:

- Was it easier to be critical or positive?

- Is it easier doing this exercise in a group? Can you build on each other?

- How easy was it to make the leap from the raw data on the grid to new innovations?

- Did you stifle any ideas or thoughts? Why?

- How could the process have been improved?

❖ Force Field Analysis ❖

🕐 *30 – 60 minutes*

Purpose

The purpose of this exercise is to give participants a tool to use for implementing an innovation or a change.

Materials

✓ Flip chart paper
✓ Markers
✓ Worksheet #21

Procedure

1. Distribute Worksheet #21 to participants.

2. This process, invented by Kurt Levin, is used widely around the world.

3. Divide the larger group into smaller groups of 4 to 7 participants.

4. Each group needs to choose an innovation or change they wish to implement in their organizations.

5. Perform a force field analysis with participants as an example. The steps are:

   a) Choose a change or an innovation (for example, smoking cessation, implementation of a new ERP system, a reorganization, a new billing system)

   b) Have the participants name a best-case scenario and worst-case scenario for the change, and write those down.

   c) Create a large "T" on a piece of flip chart paper (or use the one on Worksheet #21). Label one column "forces for" and the other "forces against" (the change).

   d) Have participants brainstorm the forces for and the forces against the innovation or change.

   e) Then have participants brainstorm ways to strengthen the forces for the change and weaken the forces against the change.

6. After the practice exercise, the small groups can do a force field analysis of their own, choosing their own innovation or change.

7. Have the small groups report out to the larger group.

Force Field Analysis (concluded)

Debrief

The force field analysis is one of the most popular "organized brainstorming" processes used in the world. It can help groups and individuals prepare for change. It can also help groups and individuals implement a change. Ask participants where this process might be useful in their organizations. Ask them if they think many impediments to changes or innovations can be anticipated. Ask them if they could use this tool for personal changes, like losing weight or stopping smoking or changing some undesirable habit.

Worksheet #21
Force Field Analysis

Best-Case Scenario:

_____

_____

_____

Worst-Case Scenario:

_____

_____

_____

FORCES FOR (+)	FORCES AGAINST (–)

Action Step #1:

_____

_____

Action Step #2:

_____

_____

Action Step #3:

_____

_____

❖ Litmus Test ❖

🕐 *30 – 60 minutes*

Purpose

The purpose of this exercise is to give participants a rudimentary tool with which they can analyze innovations.

Materials

✓ Paper
✓ Pens
✓ Worksheet #22
✓ Laptop computer with Internet access (optional)

Procedure

1. Divide the larger group into smaller groups of 4 to 7 participants.

2. Distribute Worksheet #22 to participants.

3. Explain the purpose of this exercise to participants. You should discuss each of the criteria for successful innovations on Worksheet #22.

4. Have each small group choose some innovations and analyze them using these criteria. These innovations can be process innovations, product innovations, imagined innovations, or real innovations.

5. The smaller groups can report-out to the larger group.

Debrief

You might ask the following questions:

- Do all of these criteria make sense?

- In which circumstances might one or more of these not make sense?

- What other criteria might be valuable?

Worksheet #22
Criteria for Innovations

(Some of these ideas are adapted from Joel Barker's excellent video, *Tactics of Innovation.*)

Understandable

Can potential users easily understand this innovation, or learn how to use it quickly? Board game creators often use this rubric for a new game: Does it only take less than 30 minutes to learn but years to master?

Reversibility

Can the buyer bring it back if it does not suit her/his needs? Is the downside small enough to warrant a try from new users?

Wow Factor

Is this product/service exciting or unusual? Why should people shift from their present product or service? An iPod has a "wow" factor. So does shopping at Trader Joe's, for some people.

Necessary

Does the product/service solve a problem or do an important job? Does it make life easier/better/more efficient/more joyful for its target market?

Context

Does the product have a necessary context? The Segway is a great innovation, but it does not have a necessary context, except for tours or in airports. On the sidewalk, it can run down pedestrians. On city streets, it will be run down.

❖ Attribute Listing ❖

🕐 *1 – 2 hours*

Purpose

The purpose of this exercise is to give participants an understanding of another "organized" brainstorming technique.

Materials

✓ Flip chart paper
✓ Markers
✓ Worksheet #23
✓ Wild but practical imagination

Procedure

1. Divide the larger group into smaller groups of 4 to 8 participants.

2. Distribute Worksheet #23 to participants.

3. Have each group choose a product or process that they wish to redesign. It should be something real from their workplace or something they are all familiar with, like a bike. Then they will work through the following steps to create a new version of that product or process, using the worksheet:

    • Break down the item into component parts—maybe about five of them.

    • Brainstorm five options for each of these components.

    • Choose one option from each component and create a new rendition of the original product.

4. Have the smaller groups report-out their findings to the larger group.

Debrief

Have participants note how this process has both divergent and convergent steps. Both are necessary for innovation. Ask participants if this process is more productive than unorganized brainstorming. The answer should be yes.

This process can be used later in the innovation process, after much data is gathered by research and observation. It can help put together some disparate pieces into a cohesive whole.

Worksheet #23
Attribute Listing

One way to redesign products or systems is to break them down into component parts, then look at each of the components or steps for choices or options. For instance, let us look at a baby stroller:

- Wheels
- Cover
- Handles
- Storage space
- Seat
- Axles
- Hinges
- Compactable

What options might you have for each of these component parts? List four or five for each and then collect and combine several of these to create a new baby stroller.

Think of another product. List the attributes and brainstorm alternatives for each of the attributes. Use the chart on the following page.

Product: _____

ATTRIBUTE	OPTION 1	OPTION 2	OPTION 3	OPTION 4	OPTION 5
1.					
2.					
3.					
4.					
5.					

❖ Blue Ocean Technique ❖

(This technique comes from *Blue Ocean Strategy,* written by Kim and Mauborgne, a best-selling book about innovation and strategy.)

🕐 *1 – 2 hours*

Purpose

The purpose of this exercise is to give participants the chance to practice blue ocean innovation, a technique for creating new market space for companies.

Materials

✓ Flip chart paper
✓ Markers
✓ Worksheet #24

Procedure

1. Divide the larger group into smaller groups of 4 to 8 participants.

2. Distribute Worksheet #24 to participants.

3. Explain the theory behind blue ocean strategies. Most companies compete in the red ocean, where competitive sharks are always biting at you and your offerings, drawing blood. This is a tough, competitive environment. Organizations should be innovative and create blue ocean competitive space, where there is not any competition. "Blue ocean" is competitive space where product and service offerings are so different from the competition that the market space is not bloodied.

4. One way to look at this competitive space is to work with the grid found in Worksheet #24. Have participants choose one of their market offerings (product or service) and redesign it. Have them fill in the four quadrants—create, raise, eliminate, reduce. What new attribute can they create for that offering? What attribute could they raise the level of? What could they eliminate? What could they reduce?

5. After the participants fill in their grid, have them develop their new product or service concept and then present it to the larger group.

Blue Ocean Technique (concluded)

Debrief

"Blue ocean" has become a well-used phrase in corporate America. It is synonymous with finding a new market space where there is no competition. Easier said than done. But, again, this is an organized kind of brainstorming.

You can emphasize that this technique can be used for processes, products, or services. It is similar to questioning assumptions. Why do we need this step? Why does the price need to be this high? Why does this service have to be part of the package?

Example #1

Let's look at an example: Aldi Foods. What has this chain done to be successful?

- **Eliminate:** any frills, any services
- **Reduce:** price, SKUs, loss to aging stock, floor space
- **Raise:** speed of cashiers, corn syrup instead of sugar to lower costs, sales per square foot
- **Create:** price for shopping carts, paper bags

Example #2

Let's look at another example: Trader Joe's (same owner as Aldi).

- **Eliminate:** lost shoppers, corn syrup, grumpy or indifferent employees
- **Reduce:** price, number of SKUs
- **Raise:** own brand of foods, customer service, fun people to help shoppers, team orientation of employees
- **Create:** fun in grocery shopping business, fun website

Worksheet #24
Blue Ocean Technique Grid

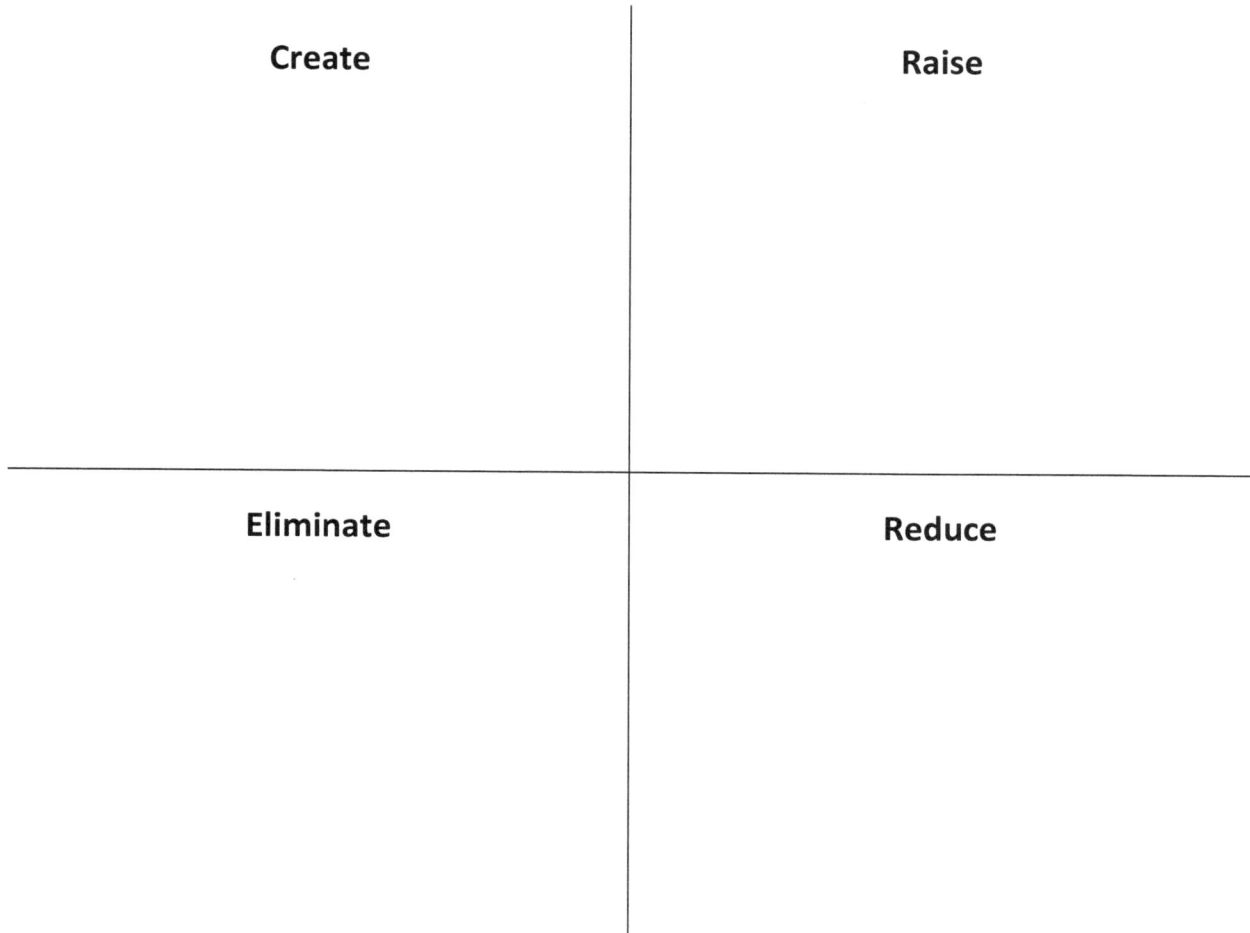

Create	Raise
Eliminate	**Reduce**

❖ Club Med Exercise ❖

🕐 *30 – 60 minutes*

Purpose

The purpose of this exercise is to give participants another organized brainstorming technique to use when innovating services.

Materials

✓ Flip chart paper
✓ Markers
✓ Worksheet #25

Procedure

1. This format has been used by Club Med, the resort company, with all of their services.

2. Divide the larger group into smaller groups of 4 to 7 participants.

3. Distribute Worksheet #25 to participants.

4. Have each group choose a process (set of steps) that their customers go through to do business with them. For example:

   a) entering, being seated, and ordering at a restaurant

   b) being billed and making a payment

   c) renting a car and returning it

5. Have the groups walk through the three steps from Worksheet #25:

   a) What are the steps in the process?

   b) At each step, what possibly could go wrong?

   c) To prevent each of these things from going wrong, what could the organization do, in terms of people, materials, and policies, to prevent them from going wrong?

6. Have the smaller groups report-out their findings to the larger group. The rest of the participants can add to their list of preventative measures.

Debrief

You can summarize the exercise by asking the following questions:

- How does this exercise relate to a standard flowchart? (It is a flowchart from the perspective of a customer.)

- Could any steps be added to improve this process?

- In some cases, could the preventative fixes be more expensive than the problems that may be created?

Worksheet #25
Club Med Exercise

What are the steps in the process?

1. _____
2. _____
3. _____
4. _____
5. _____
6. _____

At each step, what possibly could go wrong?

1. _____
2. _____
3. _____
4. _____
5. _____
6. _____

To prevent each of these things from going wrong, what could the organization do, in terms of people, materials, and policies, to prevent them from going wrong?

1. _____
2. _____
3. _____
4. _____
5. _____
6. _____

❖ Designing an Innovation Room ❖

🕐 1½ – 2 hours

Purpose

The purpose of this exercise is for participants to design a physical space where methodical innovation can happen.

Materials

- ✓ Flip chart paper
- ✓ Markers
- ✓ Notebook paper
- ✓ Pens
- ✓ Laptop computers with Internet access
- ✓ Vivid imagination
- ✓ Worksheet #26

Procedure

1. More and more companies are creating innovation rooms. These rooms provide a space for groups to brainstorm and work with ideas that they wish to bring to fruition. These rooms typically have walls made of white board material. Depending on how big the room is, there are many other items to be used in the innovation process.

2. Divide the larger group into smaller groups of 6 to 8 participants. Explain their goal—design an innovation room. If this workshop is taking place at a corporation and groups know of a room that would serve this purpose well, they might go and look at it. They also might use the Internet to find examples of innovation rooms.

3. Distribute Worksheet #26 to participants.

4. The groups might use organized brainstorming techniques like SCAMPER or Attribute Listing to perform this task.

5. Have the groups create something to present to the larger group. It may be sketches or a blueprint. It may be a PowerPoint. It may be a list of items that would be kept in bins in the innovation room (see Worksheet #26).

Debrief

You should ask some of these questions:

- What makes for a creative space?
- Can you ensure that innovation will happen?
- What materials help the creative process? Which hinder?
- What kinds of desks and chairs are best for creativity enhancement?
- What is the best way to capture all of the ideas?

Worksheet #26
Innovation Room Enhancers

- Play dough

- Pipe cleaners

- Markers, pens, crayons, scissors, other school supplies

- 3" x 5" cards

- Post-it Notes

- Bin full of broken parts
  - light fixtures
  - on/off switches
  - old toys
  - metal pieces
  - tools
  - kitchen implements
  - etc.

- Books about innovation

- Yard stick

- Poster boards

- Glue, tape, duct tape, and other adhesives

- Old magazines with pictures

- Computer projector

- Copious amounts of dark chocolate (helps with serotonin enhancement)

❖ SCAMPER ❖

🕐 *30 – 60 minutes*

Purpose

The purpose of this exercise is to introduce participants to an organized brain-storming technique entitled SCAMPER.

Materials

✓ Flip chart paper
✓ Markers
✓ A sense of humor
✓ A deck of "scamper" cards (optional)

Procedure

1. Introduce participants to the acronym SCAMPER:

   - **S**ubstitute
   - **C**ombine
   - **A**dapt
   - **M**inimize, maximize
   - **P**ut to other uses
   - **E**laborate, eliminate
   - **R**everse, rearrange

2. Explain to participants that this is an organized brainstorming technique.

3. To warm up participants, give them some funny items with which to work. Ask them what they could do with the following things to make them useful or profitable:

   - dry Christmas trees in early January
   - moldy pumpkins after Halloween
   - obsolete tanks at the end of the Cold War
   - 1,000 leftover bricks
   - 6 tons of ball bearings
   - an abandoned factory in Detroit
   - 10 retired lawyers

4. Have each group tackle several of these using the SCAMPER method, taking each step in order. Have each group develop two creative innovations for each of the items.

5. The smaller groups will report-out to the larger group and have the larger group vote on each of the innovations.

SCAMPER (concluded)

6. This can be used as a warm-up exercise. At this point, you may move the group into a real innovation exercise. For instance, ask the groups to use SCAMPER to develop innovative uses for some internal issues:

    - excess capacity in the advertising department
    - pounds of excess shipping material
    - hundreds of discarded shipping pallets
    - 300 old cubicle dividers
    - 500 volunteer hours
    - 10 retired executives who are bored

Debrief   SCAMPER is an old and tested technique. It can be used in a variety of situations. Ask participants if they came up with more ideas because of the prompts. The purpose of the scamper prompts is to spur creativity.

Posters and Quotes

50 Innovation Techniques: Addendums

Probe the Constituency	• In this step, innovators identify who the client is, what the new IT is, what the constraints are, and what research and resources are necessary to become informed. The goal of this step is to gain full understanding of the issue(s).
Observe the Real Situation	• Through observation, innovators observe real world application of the issue. The observation should be focused on what makes the customer tick and what confuses and confounds. • Mountains of data are gathered during this phase of the process.
Develop New Concepts	• Developing new concepts can be defined in one word: *brainstorming*. This step of the process calls for any and all ideas to be generated and shared, as Brynteson says, both the "outlandish" and "impractical."
Converge and Build Prototypes	• Brainstorming ideas are distilled down into workable prototypes. • The good and bad are weighed, and decisions are made.
Implementation Process	• The central question here is, "How do we put this new process/procedure into action?"

Step #1: Probe the Constituency: Questions to Consider

- How do others do the task?

- What are their constraints?

- What other technologies can be employed here?

- Name five other processes that this process is like.

- Name five other processes that this process is *not* like, but may be appropriate, with the use of a little imagination.

- What box can we step out of?

- What possible boxes can we step into?

- What is the history of this process? Has the history been linear or discontinuous?

- What has worked well? What has not?

- What is customer input?

- Who are all of the constituencies?

- What are the formal feedback mechanisms?

- In an ideal world, what would they want?

- Draw realistic pictures.

- Draw unrealistic pictures.

- What are the assumed constraints of the project?

- What other secondary data are available but not explored?

Step #2: Observe the Real Situation: Questions to Consider

- Where can you find people using this process or system?

- Can you observe them without them changing their behavior?

- What method is best for documenting what you see?
  - ✓ digital camera
  - ✓ video camera
  - ✓ copious field notes (with a laptop)
  - ✓ paper and pencil
  - ✓ other

- What kinds of questions can you use to obtain the deepest information?

- Are there other ways of obtaining information that you haven't thought of?
  - ✓ Watch users using the process.
  - ✓ Observe data trends.
  - ✓ Observe users of data.
  - ✓ What is the information used for?
  - ✓ How can this information be more accessible and more useful?
  - ✓ What confounds and confuses customers?
  - ✓ What feedback loops are in place that would make this more of a learning organization?
  - ✓ Do a data dump/synoptic learning mind map to collect and disseminate the data.
  - ✓ Take pictures.
  - ✓ Do interviews.
  - ✓ Take copious field notes.

Step #3: Develop New Concepts: Questions to Consider

- Are we stretching ourselves far enough?

- What boxes have we forgotten?

- What other boxes do we need to enter?

- What connections have not been made?

- What voices are not being heard? What voices are minimized?

- Was convergence too fast?

- What are the unintended consequences of the new prototype?

- How engaged are the participants?

- When are they engaged? Not engaged?

- Are there wacky enough ideas?

Step #4: Converge and Build Prototypes: Questions to Consider

- Is the focus in the right place?

- Can the group poke fun at their own prototypes?

- Is there an atmosphere of friendly competition?

- What are the unintended consequences of each of the prototypes?

- Use Post-it Notes to vote.

- Can we disassemble and reassemble the prototypes?

- What is at stake for each of the prototypes?

- Do we have aliens in the group?

- Are there deliberate provocateurs?

- How can we improve on it?

- In what ways might we...?

- Is everyone engaged? If not, why not?

- What questions are not being asked?

- Are all of the constituencies being represented?

- What will work best given the primary clients of this organization?

Step #5: Implementation Process: Questions to Consider

- Do a force field analysis.

- Who will lose if this innovation is accepted?

- Who will win if this innovation is accepted?

- What are organizational impediments?

- What are individual impediments?

- What action planning is useful?

- What is organizational readiness?

- What are the unintended consequences of this?

Categories of Innovation

- Process improvement ideas (lean manufacturing, Six Sigma)

- Derivative ideas (Starbucks, microloans)

- Breakthrough ideas (*Harry Potter,* space travel)

- Radical innovations (iPods, wireless)

Accident as Innovation

"Accident is the name of the greatest of all inventors."

– Mark Twain

Example #1

In 1928, Alexander Fleming left a window open next to a petri dish with a colony of bacteria. He came back the next morning. He looked through a microscope at his ruined experiment. He saw mold destroying the bacteria. He invented penicillin. The formula: accident plus acute observation.

Example #2

Percy Lebaron Spencer had 120 patents, mostly in the defense industry. One day, he walked by a magnetron—a machine used in radar. A chocolate candy bar in his pocket melted. He grabbed a handful of popcorn kernels and put them in front of the magnetron—they popped! The formula: accident plus observation plus experimentation.

Example #3

Eleven-year-old Frank Epperson left a mixture of soda powder and water; it froze to a mixing stick. Twenty years later, he decided to add some flavors, and lo and behold we had the "Eppsicles." The name needed some refinement: Popsicles. He received royalties for 60 million of them. The formula: accident plus memory plus experimentation.

"There's no use trying," Alice said; "one can't believe impossible things."

"I daresay you haven't had much practice," said the Queen. "When I was younger, I always did it for half an hour a day. Why, sometimes I've believed as many as six impossible things before breakfast."

From *Alice in Wonderland*

Not an inventor?

- The ballpoint pen was invented by a sculptor.

- The parking meter was invented by a journalist.

- The Wright brothers were bike mechanics, not aeronautical engineers.

- Kodachrome film was developed by a musician.

"There is nothing more difficult to carry out, nor more doubtful of success, nor more dangerous to handle, than to initiate a new order of things. For the reformer has enemies in all those that profit by the old order, and only lukewarm defenders in all those who would profit by the new order, this lukewarmness arising...partly from the incredulity of mankind, **who do not truly believe in anything new until they have had actual experience of it.**"

– Niccolo Machiavelli, *The Prince,* 1513

"Creativity is the ability to look at the ordinary, and see the extraordinary."

– Dewitt Jones

"Success is the ability to go from failure to failure without losing your enthusiasm."

– Winston Churchhill
(as quoted by John Peterman)

"From cradle to grave the pressure is on: BE NORMAL…The trouble with this is that corporate normalcy derives from and is dedicated to past realities and past successes. There is no room for…original thinking."

– Tom Peters

"The act of experimenting sets up a virtuous cycle of innovation; this cycle can constitute such a dominant characteristic of the organization that the ability to experiment and prototype efficiently and competently itself constitutes a competitively advantageous capability."

– Dorothy Leonard-Barton,
Wellsprings of Knowledge, p. 114

"Innovation requires a fresh way of looking at things, an understanding of people, and an entrepreneurial willingness to take risks and to work hard. An idea doesn't become an innovation until it is widely adopted and incorporated into people's daily lives. Most people resist change, so a key part of innovating is convincing other people that your idea is a good one—by enlisting their help, and, in doing so, by helping them see the usefulness of the idea."

– Art Fry, Corporate Scientist, 3M
Fast Company, April 2000, p. 100

"We have identified a 'third place.' And I really believe that sets us apart. The third place is that place that's not work or home. It's the place our customers come for refuge."

– Nancy Orsolini, District Manager
Starbucks

Innovation Cases

Navy Patrol Boats

The Problem:

It was very difficult to move heavy electronic equipment in and out of the holds of small navy patrol boats. These patrol boats were designed and built before sophisticated electronics equipment was used for tracking other ship movements. These patrol boats were being used by an Asian country to intercept gun runners and El Quaida operatives. The vertical passageways were narrow and difficult to maneuver. In moving this heavy electronic equipment, many sailors experienced back and shoulder problems.

The Solution:

A group of sailors, led by innovation consultants, studied the problem extensively. Pairs of the sailors attempted to find other "boxes" that might hold answers. For instance, one group looked into robotics for an answer. They found a robotic device on the market that climbs and descends stairs with heavy loads.

Coincidentally, a cave-in at a subway construction site in an Asian country trapped several of the workers. Emergency teams worked to free these men from under the rubble. They used compact tripods with heavy-duty ropes and pulleys that could be set up anywhere. This mechanism folded up into a duffle bag and could lift in excess of 500 pounds.

Both of these solutions evolved from "getting into another box," not reinventing the wheel. Innovation does not have to be exclusively inventing something new. The world is abundant with solutions—we just need to find the right one for the problem at hand. In this case, the tripod/ropes/pulley options were much more portable and cost-effective than high tech or robotic solutions.

Air Force: Missing Tools

The Problem:

Crews frequently maintained F-17 fighter jets for an Asian Air Force. If, after the jets took off on a mission, a tool could not be found, the jets were summoned back to the airbase, often at great expense. The fear was that the tool might have been left in the engine. A tool rattling around in an engine might bring the plane down in a crash.

The current inventory system for tools was simple yet cumbersome. Tools were checked out by hand and entered in a log system in long hand. It often took 20 minutes. Likewise when the tools were turned in, they were logged in, long hand. It became more complex when multiple tools were checked out quickly or checked back in quickly. One might be left out on either side of the process. If so, a decision had to be made on whether to scramble the aircraft back to the airbase.

The Solution:

A team of airmen, led by innovation consultants, created a check-in, check-out system, with radio-frequency identification (RFID) tags, that made tracking the tools much simpler and time-effective. In order to arrive at this solution, the airmen had to look at parallel processes. They examined library systems and package tracking systems. Although this system required an upfront investment, it could be used for other issues at this airbase and was adopted to do so.

Infantry: Night Vision Goggles

The Problem:

Very expensive night vision goggles were disappearing from the infantry of an Asian country. Some feared that they had fallen into El Quaida hands. Others thought they were being sold by underpaid soldiers on the black market. Others believed that they had just fallen off soldiers and had been lost in swamps during thick jungle maneuvers. Whatever the case, these precision goggles were too expensive to be lost on a regular basis.

The Solution:

An infantry team, facilitated by innovation consultants, thoroughly investigated the issue. One of their first actions was to have a video taken of an infantryman affixing the goggles to his helmet. The process took almost eight minutes as they used hard plastic twist ties to attach the goggles and then trimmed the ends. The team was appalled as they watched the video several times. Note that this was the best way for the team to get into the real situation—they needed to observe the entire process in order to understand it.

The team then broke down the problem into component parts: storing the device, attaching the device, and finding the device when lost. For the first issue, the team consulted a seamstress for a redesign on the infantryman's vest. She helped create a robust pocket with a Velcro and snapping system to ensure that the goggles would not fall out.

The team turned its attention to the attachment problem. The team, with the help of an engineer, devised a clipping system for the helmet. Although this solution required a clipping system for both the helmet and the goggles, it was felt that the incremental expense was well worth the additional security.

Finally, the team examined the issue of lost goggles. It looked at several honing devices and flashing light devices. This issue was complicated by the fact that the devices should not be able to be seen by the enemy at night or during the day. Many of these tracking devices could possibly give away troop movement and locations. The team settled on an RFID tracking system with small chips embedded in the helmets.

In the case of these disappearing goggles, breaking the problem into component parts (similar to Attribute Listing) provided the best process for innovation. Each of the three components needed different innovative approaches.

Infantry: Night Vision Goggles

The Problem:

Infantry personnel in a small Asian country performed operations in the jungle at 90-degree temperatures and 90 percent humidity. During the course of these operations, several soldiers passed away from carrying too much weight in these difficult conditions. In particular, those who died were carrying very heavy rocket propelled grenade launchers and the rockets for them.

One of the issues faced by the innovation team was the nature of the real problem. Was the real problem how to distribute the weight of all of this equipment or was it that there was too much weight? Some research indicated that men under these conditions cannot possibly carry that much weight.

The Solution:

A team of infantrymen and innovation consultants, with the help of an ergonomics expert and a backpack manufacturer, worked together to find a safer way to carry the weight. They created a backpack and strap system that made carrying large amounts of weight easier.

Live Fire Zone

The Problem:

In live ammunitions night maneuvers, infantrymen could not determine whether they hit their targets. If they did not hit the target, they could not move ahead to the next target. An ancillary problem was that some of the expensive targets were destroyed by 50mm machine gun fire.

A team from the officer training school and an innovation consultant studied the problem by immersing themselves in the real situation. They walked the live fire zone, studied the targets, and observed the maneuvers. They studied how the most sophisticated armies of the world dealt with this problem.

The Solution:

The team discovered a 3M reflective tape that could be used on low-tech targets and that glowed in the dark if hit. Low-tech targets were not damaged by 50mm bullets. Mission accomplished: the team saved money on the targets and found an effective way for soldiers to discern whether they had hit the targets.

Chinook Helicopters: Hydraulic System

The Problem:

Chinooks are large helicopters used for carrying heavy equipment like jeeps and armored cars. They can land equipment in battle zones quickly for fast deployment. The testing time for one small army for the hydraulic systems of Chinook helicopters was 3.5 hours, far too long. In other words, the hydraulic system that dropped a heavy chain and picked up the cargo needed to be tested frequently. A three and a half hour turnaround time for the tests was unrealistic in wartime.

The Solution:

A military team, led by innovation consultants, decided to try to "get into other boxes." In teams of two, they studied other types of hydraulic systems, either in person or through www.howthingswork.com. They examined automobile hydraulic systems, elevator hydraulics, and the cranes that unloaded containers from freight ships. Their investigations indicated that there were many ways to test hydraulics. They found one process that was simple and effective. It ultimately took only 30 minutes to test the hydraulics of these helicopters.

A New Television Program

The Problem:

A major television network needed some new hit shows.

The Solution:

A team of producers and other creative types from the network gathered with an innovation consultant at a resort for two days. Some of the creative processes that the group undertook included the following:

- The group conducted a brainstorming session based on data and statistics in order to determine which demographics were the prime targets for new shows. This dive into the data yielded tweens, young professionals, seniors, and stay-at-home mothers.

- The group was split into four smaller groups, each representing one of those demographics. Each small group did a "25 questions," where they delineated 25 questions that they have about that demographic. They then researched those questions, using the Internet, public documents, and previous market research conducted by the network. The groups then did a data dump around their demographics.

- Each group then developed three to four scenarios for new programs based on their data dump. They weighed the pros and cons of each possibility. After evaluating each option, each group chose one program option.

- Each group developed the main characters and plot lines, mapped the first three episodes for their program, and created a set of storyboards to present to the rest of the group.

Notice that this two-day ideation session combined divergent and convergent thinking to capture optimum solutions.

New Corporate Processes

The Problem:

A large multinational computer company had opened a new facility in an Asian country. It was adding new personnel quickly and needed to add more. Because of the fast growth, the culture was in flux and not developing the depth of the corporate culture at headquarters.

The Solution:

A group of managers and an innovation consultant broke the culture problem into four manageable components:

- Recruiting 600 new highly skilled people

- Developing an onboarding process

- Creating a happy, fulfilling workplace

- Developing an innovative culture

Teams took ownership of each of these problems. Teams decided to dig deeply into each of these issues by examining what best-in-class companies do about these issues. Members of the teams interviewed employees in other high-tech companies. They researched the Internet and read case studies.

Each team developed a multitude of initiatives to solve their problem. The initiatives were evaluated for feasibility and then ranked. Some were put on the front burner. Others were put on the back burner. Action plans were created for the front burner items.

References

Amabile, T. M. (1998). "How to kill creativity." *Harvard Business Review*. (September/October): 77–87.

Barker, J. (2001). *The business of paradigms.* St. Paul, MN: Star Thrower Distribution.

Barker, J. (2005). *Five regions of the future.* London: Penguin.

Barker, J. (2009). *Innovation at the verge.* St. Paul, MN: Star Thrower Distribution.

"Best innovations of 2007." (November 12, 2007). *Time Magazine.*

Bughin, J., Chui, M., & Johnson, B. (2008). "The next step in innovation." McKinsey Quarterly.

Carter, S. (1999). *Renaissance management: The rebirth of energy and innovation in people and organizations.* London: Kogan Page Limited.

Chowder, K. (September 2003). "Eureka." *Smithsonian Magazine,* p. 92.

Christensen, C. M. (1997). *The innovator's dilemma: When new technologies cause great firms to fail.* Boston: Harvard Business School Press.

Collins, J. (2001). *From good to great.* New York: HarperBusiness.

Drucker, P. (1985). *Innovation and entrepreneurship: Practice and principles.* New York: Harper and Row.

Drucker, P. (August 2002). "The Discipline of Business." *Harvard Business Review.*

Freidel, R. (October 1996). "The accidental inventor." *Discover Magazine,* p. 69.

Fry, A. (April 2000). *Fast Company Magazine.*

Gelb, M. (2007). *Innovate like Edison.* New York: Dutton.

Goleman, D. (1995). *Emotional intelligence.* New York: Bantam.

Gryskiewicz, S. S. (1999). *Positive turbulence: Developing climates for creativity, innovation and renewal.* San Francisco, CA: Jossey-Bass.

Hamel, G. (February 2006). "The why, what, and how of management innovation." *Harvard Business Review,* pp. 72–84.

Hargadon, A., & Sutton, R.I. (May-June 2000). "Building an innovation factory." *Harvard Business Review.*

Hargadon, A. (2003). *How breakthroughs happen: The surprising truth about how companies innovate.* Boston: Harvard Business School Press.

Heath, C., & Heath, D. (2007). *Made to stick.* New York: Random House.

Kim, W. C., & Mauborgne, R. (2005). Cambridge: Harvard Business School Press.

Leonard, D., & Rayport, J. F. (November–December 1997). "Spark innovation through empathetic design." *Harvard Business Review.*

Leonard-Barton, D. (1995). *Wellsprings of knowledge: Building and sustaining the sources of innovation.* Boston: Harvard Business School Press.

McKinsey and Co. (2007). "How companies approach innovation." 2007 McKinsey on Innovation.

McKinsey and Co. (2010). "Open source innovation" (working paper).

Michalko, M. (1991). *Thinkertoys.* Berkeley: Ten Speed Press.

Miller, W. C. (1998). *Flash of brilliance: Inspiring creativity where you work.* Reading, MA: Perseus Books.

Pink, D. (2005). *A whole new mind.* New York: Riverhead Books.

"Powers of creation." (October 1996). *Discover Magazine.*

Schrage, M. (1989). *No more teams.* New York: Doubleday.

Senge, P. (1990). *The fifth discipline.* New York: Currency Doubleday.

Smith, F. (April 2000). *Fast Company.*

Stacey, R. D. (1996). *Complexity and creativity in organizations.* San Francisco, CA: Berrett-Koehler Publishers Inc.

Tapscott D., & Williams, A. (2010). *Wikinomics: How mass collaboration changes everything.*

"The power of invention." (Winter 1997–1998). *Newsweek Extra.*

Weick, K. E. (1996). "Drop your tools: An allegory for organizational studies." *Administrative Science Quarterly,* p. 2.